Insect Ecology

The Authors

Dr. Shadanan Upadhyay born in 24 October, 1956. He took his M.Sc. Ag. Entomology and Ph,D. Degree from J.N. Krishi Vishwa Vidyalaya, Jabalpur during 1979 and as in service candidate in 1991, respectively. He was posted as Junior Scientist in sugarcane research station Jaora, district Ratlam of Madhya Pradesh. Presently he is posted as professor and head of the section, department of entomology, College of Agriculture, Indore, M.P. He has more than 35 years of experience of teaching, research and extension. He has published more than 35 research papers in repute of journals. He has also published several extension articles, bulletins etc.

Dr. S.B. Singh born in 1st July, 1963 in Rewa district of Madhya Pradesh. He passed his M.Sc. Ag. Entomology in 1984 and Ph.D. Degree in 1995 from J.N. Krishi Vishwa Vidyalaya, Jabalpur. He was posted as research scholar in 1989 in AICRP on Honey Bee (Res. And Training) in College of Agriculture, Indore, Madhya Pradesh. Presently he is posted as Assistant Professor (Entomology), Department of Entomology, College of Agriculture, Indore, M.P. He has gained wide experience of teaching, research and extension in his 26 years of services. He has published more than 20 research papers in repute of journals and a number of abstracts in many seminars and symposia. He has written seven manuals on different aspects of teaching. He has also several extension articles and bulletins publication.

Dr. Satya Prakash Singh Tomar working as a Scientist, Ag. Entomology (Plant protection) and Assistant Registrar (Academic/ Legal) in Rajmat Vijayaraje Scindia Krishi Vishwa Vidyalaya, Gwalior. He was Born on 6 August 1979 in Chambal region of M.P., he did M.Sc. (Agric.) from JNKVV, Jabalpur (2003); Ph.D. (Zoology with specialization in Ag. Entomology) from Dr. B.R. Ambedkar university, Agra (2009);. He joined as Subject Matter Specialist, JNKVV, Jabalpur (2007). Thereafter he served as Scientist in many agro climatic zones of Madhya Pradesh.

Dr. Tomar has more than 200 publications in the form of research papers, books/ bulletins, seminar/symposia proceedings, popular articles, manuals and technical reports on various aspects of Entomology, Insecticide Resistance Management, IPM, Plant protection, Beekeeping and Honey processing & packaging. Out of these, 150 research papers are in Peer reviewed journals.He published

He received Award of honour 2012 was given by M.P khadi Gramodyog sangh, Bhopal (M.P.) for the work on development of employment among the rural youth in Morena district through beekeeping training, SHGs training, Honey processing and marketing related activity; Young Scientist Award (2007); Best writer Award (2005, 2007, 2008), and Best Research Worker Award (2005-2007), He was invited to present lead or scientific papers at international conferences and in several national conferences.

Insect Ecology

— *Authors* —

Dr. Shadanan Upadhyay

Professor (Entomology)

Dr. S.B. Singh

Assistant Professor (Entomology)

Dr. Satya Prakash Singh Tomar

Scientist (Entomology)

Rajmata Vijayaraje Sciendia Krishi Vishwa Vidyalaya
Gwalior, M.P., India

BIOTECH

2021

BIOTECH BOOKS®

ISBN 978-81-7622-373-7

Published by: **BIOTECH BOOKS®**
4762-63/23, Ansari Road, Darya Ganj,
New Delhi - 110 002
Phone: +91-011-23262132
E-mail: biotechbooks@yahoo.co.in

Printed at: **Replika Press Pvt. Ltd.**

PRINTED IN INDIA

Preface

Effective management of ecosystem resources depends on evaluation of the complex, often complementary, effects of insects on ecosystem conditions as well as insect responses to changing conditions. Two emerging needs require the integration of traditional and emerging mental changes that must be addressed from a global perspective, with emphasis on integrating ecological processes at various levels of resolution. Insect population structure, interaction with other species, and effects on ecosystem processes are integral to explaining and mitigating global changes. The present book is a compilation of various chapters including Insect Ecology: An introduction, living and nonliving components, environmental ecological factors, climatic factors, water or moisture, temperature, light, topographic factors, biotic, edaphic factors, ecosystems etc. This book is a sincere attempt to solve the problems of the students and faculties attempting to learn & teach insect ecology respectively. Non availability of all the contents in a single book has incited us to collect the relevant matter and compile them into a single book designed especially for the students of ecology. Our main motive is to cover the syllabi of varied Universities as far as possible. The book entitled "Insect Ecology" aims at clarity so that the text can be selective and meaningful. This text demonstrates how evolutionary and ecosystem approaches complement each other, and is intended to stimulate further integration of these approaches in experiments that address insect roles in ecosystems.

We hope that the general Entomologists and students of Agricultural entomology will find the book valuable & beneficial regarding information about the subject as well as allied branches. The book provides information, chapter wise as per ICAR course curriculum of PG Entomology programme. We are especially grateful to our colleagues at various Institutions and Universities for their help rendered for the designing of this text. We owe our sincere thanks to the publisher for his whole hearted support.

Dr. Shadanan Upadhyay

Dr. S.B. Singh

Dr. Satya Prakash Singh Tomar

Contents

Preface v

1. History, Definitions and Basic Concept 1

2. Organization of the Biological World 7

3. Abundance, Distribution and Diversity of Insects 11

4. Basic Principles of Abiotic Factors and their Generalized Action on Insects 23

5. Implications for Abundance and Distribution of Organisms including Insects: Law of Minimum, Law of Tolerance and Biocoenosis, Systems Approach to Ecology 31

6. Basic Concepts of Abundance, Model v/s Real World, Population Growth Basic Models, Exponential v/s Logistic Models, Discrete v/s Continuous Growth Model 37

7. Vital Statistics: Life Table and their Application in Insect Biology 47

8. Population Dynamics – Factors Affecting Abundance – Environmental Factors, Dispersal and Migration, Seasonality in Insects, Classification and Mechanism of Achieving different Seasonality – Diapauses (Quiescence) – Aestivation, Hibernation 57

9. Biotic Factors: Food is a Limiting Factors for Distribution and Abundance, Nutritional Ecology 75

10. Food Chain: Web and Ecological Succession 81

11. Interspecific Competition: Lotka-Volterra-Model 101

12. Concept of Niche: Ecological Homologous Competitive Exclusion Prey-Predators Interactions – Defence Mechanism against Predators/Parasitoids – Evolution of Mimicry, Colouration, Defense Mechanism against Predators/Parasitoids **109**

13. Community Ecology: Concept of Guild, Organisation of Community **129**

14. Relative Distribution of Organism, Concept of Diversity: The Wallacian View **133**

15. Assessment of Diversity, Diversity Stability, Debate Relevance to Pest Management **143**

16. Pest Management as Applied Ecology **167**

Index **181**

Chapter 1

History, Definitions and Basic Concept

History

Ecology is a new science, having only become prominent in the second half of the 20th Century. Nonetheless ecological thinking at some level has been around for a long time, and the principles of ecology have the developed gradually, closely intervened with the development of other biological disciplines.In the 19th century many ecologists like Clements (1905), Warming (1909) have described ecology. Adams (1906, 1909) described various animal communities of Northen Michigan and Isle Royale. Shelford (1912 and1919) gave examples of ecological succession, animal communities and effects of various environmental factors. He gave importance to physiological ecology. L.R. Dice (1943) and F. F. Clements (1939) gave the concept of biotic principles. The important works are as follows:

Year	*Scientist*	*Contribution*
1735	Carl Linnaeus	Swedish naturalist developed his system for classifying and naming plants and animals.
1798	Thomas Malthus	English reverend produced the earliest theoretical study of population dynamics.
1822	Gregor Mendel	Studied the inheritance of traits and led to the development of population Genetics.
1838	P. F. Verhulst	Developed formula for population growth when resources are limiting.
1840	J. Leibig	Studied limiting factor and the limited supplies of nutrients in growth and development of plants.
1859	Charles Darwin	English naturalist published his theory on the Origin on the species.

Contd...

Contd...

Year	Scientist	Contribution
1869	Ernst Haeckel	Germen Zoologist first defined the term ecology.
1877	Karl Mobius	Germen zoologist publishes paper on oyster beds and uses the word "biocoenosis" As a descriptor of an animal community; may be first use of "community"In modern ecological sense; also develops idea of community equilibrium.
1887	Stephen A. Forbes	Publishes "The Lake as a microcosm" (1887) foundation of modern limnology; self educated zoologist, published many other key ecology papers before 1900.
1895	F.A.Forel	A Swiss coined and founds science of limnology.
1899	Henry Cowles	US botanist published his classic paper on succession in sand dunes on Lake Michigan, USA.
Early 1900s	Victor Shelford and C.C. Adams	Contemporaries of Cowles, looked at animal communities in the same manner that plant ecologist did, classification and dynamics, Shelford is first president of Ecological society of America, established in1915;
1905	F.F. Blackman	Factors interaction *e.g.* light,$CO_{2,}$ assimilation rate etc.
1908	G.H.Hardy and W. Weinberg	Independently wrote the classic papers on genetic equilibrium in populations that make the beginning of population genetics.
1913	Shelford	Published book in 1913 on Animal communities of the Chicago area. He emphasised the interaction of plants and animals.
1913	V.E. Shelford	Animal communities in temperate America. He developed the law of tolerance, which linked and organisms to its environment.
1913	V.E. Shelford	British Ecological Society founded
1915	V.E. Shelford	British Ecological Society founded
1916	V.E. Shelford	US Ecologist Frederic Clements coined the phrase invertebrate "climax communities" for large areas of rather uniform vegetation which he attributed to climatic factors.
1925	Alfred J. Lotka	Publishes the elements of physical Biology includes laws of thermo dynamics along with Vito Volterra , an Italian mathematician and Raymond Pearl devise basic equation later widely used in population and community ecology (1920,s = 'golden age of mathematical population ecology'). They studied population growth under limiting conditions, predation and competition. Their work established the foundation population ecology concerned with population growth regulation and intra specific and inter specific competition.
1926	August Thienemann	Introduced the idea of organic nutrient cycling and feeding levels, using the terms producers and consumers.
1926	NIVavilov	Russian botanist published centres of origin of cultivated plants, concluding that there are relatively few such centres, many of which are located in mountainous areas.
1927	C Elton	Published Animal ecology, emphasizing regulation of population size, ecological niches and food chains.
1930	H.Stoddard	Gave concept of applied ecology with the work on the ecology of fire.
1930	A. Leopold	Worked on the application of ecological principles to wild life management.

Contd...

Contd...

Year	Scientist	Contribution
1934	G F Gause	Russian ecologist first stated the principles of competitive exclusion, related to a species niche.
1935	Arthur Tansley	British ecologist first coined the term ecosystem.
1939	R. Hesse	Published - Ecological Animal Geography
1940	Charles Eltan	Population biologist developed the idea of trophic levels in a community of organisms.
1949	Allee, Emerson, Park, Park and Schmidt	Published the comprehensive book named principles of Animal ecology (1949). It emphasized energy budgets, population dynamics, evolution and natural selection
1950	Allee, Emerson, Park, Park and Schmidt	The theory that natural selection may favour either individuals with high reproductive rates and rapid development(r- selection) or individuals with low reproductive rates and better competitive ability (k-selection) is first discussed.
1957	G.E. Hutchinson	Limnologist and Zoologist, publishes many very influential books and papers on niche, energy flow and nutrient budget.
1979	James Lovelock	English naturalist proposed his Gala hypothesis, viewing the planet as a single organism.
1980	National Science Foundation (NSF)	NSF began the program called long term ecological research (LTER)
1993		UN Convention on Biological Diversity came into force.

Definitions

Ecology

It is the science concerned with the relationships between living organisms and their environment; derived from oikos meaning house and logos -to study.

Habitat

Physical surroundings, in which an organism lives- consists of various physical attributes such as climate, soil types, moisture, but also competition among species.

Niche

The role or function played by an organism in its ecosystem; its activities, resource use and interactions with other organisms.

Community

A grouping of organisms within a shared habitat includes plants and animals (biomass) refer to the shared system. Communities are not self sufficient, often using or requiring inputs from other communities. The community and the abiotic components of an environment comprise an ecosystem.

Ecosystem

A community of organisms and their interactions within the physical environment; Energy and matter "flow" through an ecosystem through the food chain.

Evolution

It is the process in which species are created and refined through natural selection.

Natural Selection

Nearly all individuals display some variation. Variations that lead to competitive advantages will tend to allow the individual to produce more offspring and will tend to spread through the species. Variation occurs naturally (DNA) and is promoted when the genes of two individuals each with their own variations are mixed.

Living things evolve over time to adapt to changes in their environment or to take advantage of an opportunity in the ecosystem. A process called natural selection selects these adaptations. The process of natural selection is often called Darwin's Theory. Charles Darwin wrote a book called **"The Origin of Species"** in which he theorized that present forms of life on earth have descended from previous forms of life with certain changes. He believed these changes were selected for by nature in that they allowed the species to be more successful in some fashion.

Darwin observed that:

1. In any population, members of the population all display some variation from one another.
2. Many more organisms are born than survive to reproduce
3. Resources are limited and members of a population compete for survival
4. Survivors live to reproduce and pass on their genetic code to the next generation.

In this way the individuals best adapted to the environment or best able to compete reproduce more often and by inference more successfully and pass on the trait that allowed them to compete, hence nature selects them over a less successful competitor. The minute variations from individual to individual are small incremental mutations, which are chemical changes in the DNA that result in a different expression. These changes are persistent, that is they can be inherited.

In order to perform the work of living, an organism needs to extract energy from its environment. The chemical processes used by living organisms to conduct the biochemical reactions that are life use energy. Energy cannot simply be created by the organism; organisms convert energy from one form to another by means of their metabolisms, which is through controlled chemical reactions.

Basic Concepts

The study of animal ecology is probably the oldest science and has formed the foundation of much of mankind's exploitation of the natural world. The study has ranged from identification of edible plants and animals to more complex and subtle studies of the interactions between organisms and their physical, chemical and biological environments. If we are to make sustainable use of our environment and cater to the needs of the society (produce food, clothing, shelter and fuel, reliably and efficiently), we must study and understand the ecology of both the natural world and the man-made ecosystems that result from our use of this natural world. As we intensify the demands on our environment, agriculture has become ever more important for feeding and clothing us. Information about insect ecology is sought for a variety of purposes. Ecological studies of potential pests have become central to the understanding necessary for us to meet the challenges of obtaining sufficient protection of our crops and livestock.

In nature, insect species exist and evolve as components of 'communities' of plants and animals in particular habitats. Some insect species, including many plant feeders, exist only in one type of community; others, including many predators, exist as members of a number of community types, which may differ in floral and faunal composition. All insect species have a limited distribution range; and, characteristically, insect numbers fluctuate to a greater or lesser extent both in time and in space.

The word ecology is derived from Greek words *eikos* (= house or habitation) and *logos* (= study or discourse). The term was first coined by Haeckel (1869) and is defined as the total relations of the living organisms to both its organic and inorganic environment. It is also known as the study of the interrelation of organisms or group of organisms to their environment. Ecology has also been termed as *environmental biology*.

- ☆ **Odum (1963)** defined ecology as the study and function of nature.
- ☆ **Huxley and Wells (1939)** described ecology as an extension of economics to the whole world of life.
- ☆ **Elton (1927)** termed it as the study of biotic communities and science of community populations.
- ☆ **Andrewartha (1961)** defined it as the scientific study of the distribution and abundance of organisms.
- ☆ **Kendeigh (1961)** defined ecology as the study of animals and plants in relation to each other and their environment.

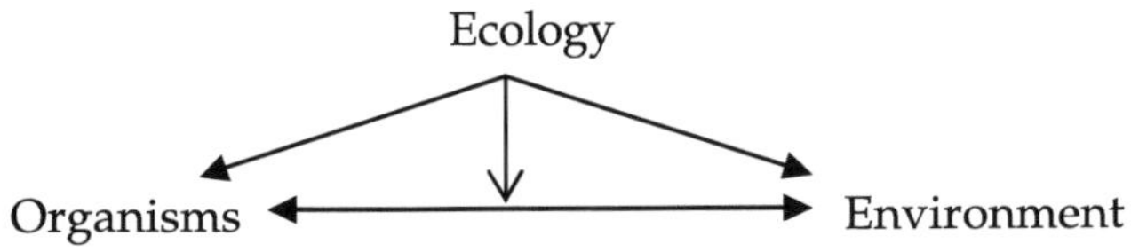

Modern Concepts of Ecology

Accordingly, ecology is considered in terms of the concept of levels of organization, which constitutes the biological spectrum (comprising – biotic and abiotic components and their resultant bio-systems). The **molding** of the population by means of natural selection is one of the basic concepts in ecology. Another fundamental concept in ecology is that energy flows through the ecosystem, and this is contrasted with the cycling of chemicals in the system. Thinking ecologically enables a holistic concept of the organism in the ecosystem.

Subdivisions of Ecology

- ✰ Animal ecology and Plant ecology
- ✰ Autecology (individuals) and Synecology (groups)
- ✰ Habitat ecology
- ✰ Paleoecology
- ✰ Cytoecology
- ✰ Conservation ecology
- ✰ Production ecology

Chapter 2
Organization of the Biological World

Biological organization, or the **hierarchy of life**, is the hierarchy of complex biological structures and systems that define life using a reductionistic approach. The traditional hierarchy, as detailed below, extends from atoms (or lower) to biospheres. The higher levels of this scheme are often referred to as **ecological organization**.

Each level in the hierarchy represents an increase in organizational complexity, with each "object" being primarily composed of the previous level's basic unit. The basic principle behind the organization is the concept of *emergence*–the properties and functions found at a hierarchical level are not present and irrelevant at the lower levels.

Organization further more refers to the high degree of order of an organism (in comparison to general objects). Ideally, individual organisms of the same species have the same arrangement of the same structures. For example, the typical human has a torso with two legs at the bottom and two arms on the sides and a head on top. It is extremely rare (and usually impossible, due to physiological and biomechanical factors) to find a human that has all of these structures but in a different arrangement.

The biological organization of life is a fundamental premise for numerous areas of scientific research, particularly in the medical sciences. Without this necessary degree of organization, it would be much more difficult–and likely impossible–to apply the study of the effects of various physical and chemical phenomena to diseases and body function. For example, fields such as cognitive and behavioural neuroscience could not exist if the brain was not composed of specific types of cells, and the basic concepts of pharmacology could not exist if it was not known that a change at the cellular level can affect an entire organism. These applications extend into the ecological levels as well. For example, DDT's direct effect occurs at the subcellular level, but affects higher levels up to and including multiple ecosystems. Theoretically, a change in one atom changes the entire biosphere.

The simple standard biological organization scheme, from the lowest level to the highest level, is as follows:

- The basic unit of all matter, the atom
- The molecule, a grouping of atoms
- The organelle, a functional grouping of biomolecules
- The cell, the basic unit of all life and the grouping of organelles
- The tissue, a functional grouping of cells
- The organ, a functional grouping of tissues
- The organ system, a functional grouping of organs
- The organism, the basic living system, a functional grouping of the lower-level components, including at least one cell
- The population, a grouping of organisms of the same species
- The biocoenosis or community, an interspecific grouping of populations
- The ecosystem, a grouping of organisms from all biological domains in conjunction with the physical (abiotic) environment
- The biosphere, the complete set of ecosystems

More complex schemes incorporate many more levels. For example, a molecule can be viewed as a grouping of elements, and an atom can be further divided into subatomic particles (these levels are outside the scope of biological organisation). Each level can also be broken down into its own hierarchy, and specific types of these biological objects can have their own hierarchical scheme. For example, genomes can be further subdivided into a hierarchy of genes.

Each level in the hierarchy can be described by its lower levels. For example, the organism may be described at any of its component levels, including the atomic, molecular, cellular, histological (tissue), organ and organ system levels. Furthermore, at every level of the hierarchy, new functions necessary for the control of life appear. These new roles are not functions that the lower level components are capable of; they are referred to *emergent properties*.

Every organism is organized, though not necessarily to the same degree. An organism can not be organized at the histological (tissue) level if it is not composed of tissues in the first place.

Underlying Philosophy

This hierarchical scheme can be viewed as an implementation of the epistemological theory of reductionism, that a complex structure is nothing more than the sum of its parts. This philosophy does allow for the observed emergent properties, characteristics not seen at the previous level. In this way, the organization is often erroneously viewed through the theory of holism, that the complex structure can not be described by the sum of its parts. However, these emergent properties of life still fall within the realm of reductionism because, while a different function may emerge at each level of complexity, these functions can be directly explained by the processes by which the object is formed and the arrangements and interactions of the units. For example, the dissected brain is simply a network of billions of neurons that come together to produce phenomena such as cognition, sensation and regulation. The reductionist approach, and the purpose of this hierarchy, views these phenomena as a direct result of *how* these neurons are arranged and interact.

Limiting Factors

A factor or limiting resource is a factor that controls a process, such as organism growth or species population, size, or distribution. The availability of food, predation pressure, or availability of shelter is examples of factors that could be limiting for an organism. An example of a limiting factor is sunlight in the rainforest, where growth is limited to all plants in the under story unless more light becomes available (such as in the event of a tree fall).

Chapter 3

Abundance, Distribution and Diversity of Insects

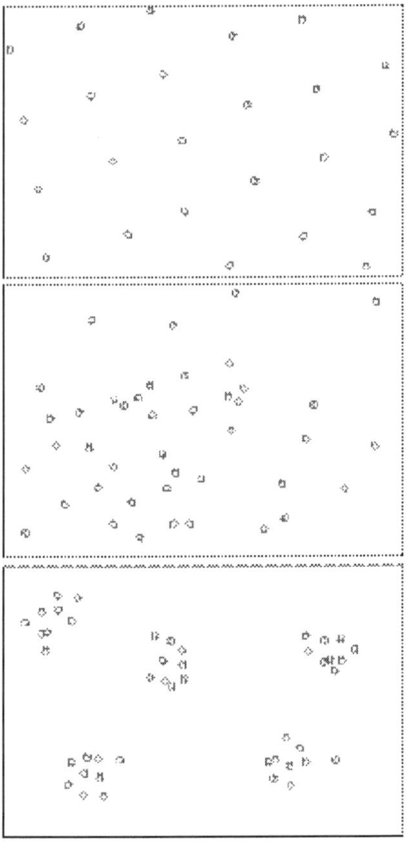

Species distribution is the manner in which a biological taxon is spatially arranged. Species distribution is not to be confused with dispersal, which is the movement of individuals away from their area of origin or from centers of high population density. A similar concept is the **species range**. A species range is often represented with a species range map. Bio-geographers try to understand the factors determining a species' distribution. The pattern of distribution is not permanent for each species. Distribution patterns can change seasonally, in response to the availability of resources, and also depending on the scale at which they are viewed. Dispersion usually takes place at the time of reproduction. Populations within a species are translocated through many methods, including dispersal by people, wind, water and animals. People are one of the largest distributors due to the current trends in globalization and the expanse of the transportation industry. For example, large tankers often fill their ballasts with water at one port and empty them in another, causing a wider distribution of aquatic species.

There are three basic types of population distribution within an area. From top to bottom: uniform, random and clumped.

Biogeography is the study of the distribution of biodiversity over space and time. It is very useful in understanding species distribution through factors such as speciation, extinction, continental drift, and glaciation, variation of sea levels, river capture and available resources. This branch of study not only gives a description of the species distribution, but also a geographical explanation for the distribution of particular species. The traditional biogeographic regions were first modeled by Alfred Wallace in *The Geographical Distribution of Animals* (1876). These were based on the work of Sclater's terrestrial biogeographic regions. Wallace's system was based on both birds and vertebrates, including non-flying mammals, which better reflect the natural divisions of the Earth due to their limited dispersal abilities.

Clumped distribution is the most common type of dispersion found in nature. In clumped distribution, the distance between neighbouring individuals is minimized. This type of distribution is found in an environment that is characterized by patchy resources. Clumped distribution is the most common type of dispersion found in nature because animals need certain resources to survive, and when these resources become rare during certain parts of the year animals tend to "clump" together around these crucial resources. Individuals might be clustered together in an area due to social factors such as selfish herds and family groups. Organisms that usually serve as prey form clumped distributions in areas where they can hide and detect predators easily.

Other causes of clumped distributions are the inability of offspring to independently move from their habitat. This is seen in juvenile animals that are immobile and strongly dependent upon parental care. For example, the bald eagle's nest of eaglets exhibits a clumped species distribution because all the offspring are in a small subset of a survey area before they learn to fly. Clumped distribution can be beneficial to the individuals in that group. However, in some herbivore cases, such as cows and wild beasts, the vegetation around them can suffer, especially if animals target one plane in particular.

Clumped distribution in species acts as a mechanism against predation as well as an efficient mechanism to trap or corner prey. African wild dogs, *Lycaon pictus*, use the technique of communal hunting to increase their success rate at catching prey. It has been shown that larger packs of African wild dogs tend to have a greater number of successful kills. A prime example of clumped distribution due to patchy resources is the wildlife in Africa during the dry season; lions, hyenas, giraffes, elephants, gazelles, and many more animals are clumped by small water sources that are present in the severe dry season. It has also been observed that extinct and threatened species are more likely to be clumped in their distribution on a phylogeny. The reasoning behind this is that they share traits that increase vulnerability to extinction because related taxa are often located within the same broad geographical or habitat types where human-induced threats are concentrated. Using recently developed complete phylogenies for mammalian carnivores and primates it has been shown that the majority of instances threatened species are far from randomly distributed among taxa and phylogenetic clades and display clumped distribution.

Less common than clumped distribution, uniform distribution, also known as even distribution, is evenly spaced. Uniform distributions are found in populations in which the distance between neighboring individuals is maximized. The need to maximize the space between individuals generally arises from competition for a resource such as moisture or nutrients, or as a result of direct social interactions between individuals within the population, such as territoriality. For example, penguins often exhibit uniform spacing by aggressively defending their territory among their neighbors. Plants also exhibit uniform distributions, like the creosote bushes in the southwestern region of the United States. *Salvia leucophylla* is a species in California that naturally grows in uniform spacing. This flower releases chemicals called terpenes which inhibit the growth of other plants around it and results in uniform distribution. This is an example of allelopathy, which is the release of chemicals from plant parts by leaching, root exudation, volatilization, residue decomposition and other processes. Allelopathy can have beneficial, harmful, or neutral effects on surrounding organisms. Some allelochemicals even have selective affects on surrounding organisms; for example, the tree species *Leucaena leucocephala* exudes a chemical that inhibits the growth of other plants but not those of its own species, and thus can affect the distribution of specific rival species. Allelopathy usually results in uniform distributions, and its potential to suppress weeds is being researched. Farming and agricultural practices often create uniform distribution in areas where it would not previously exist, for example, orange trees growing in rows on a plantation. Random distribution, also known as unpredictable spacing is the least common form of distribution in nature and occurs when the members of a given species are found in homogeneous environments in which the position of each individual is independent of the other individuals: they neither attract nor repel one another. Random distribution is rare in nature as biotic factors, such as the interactions with neighboring individuals, and abiotic factors, such as climate or soil conditions, generally cause organisms to be either clustered or spread apart. Random distribution usually occurs in habitats where environmental conditions and resources are consistent. This pattern of dispersion is characterized by the lack of any strong social interactions between species. For example: when dandelion seeds are dispersed by wind, random distribution will often occur as the seedlings land in random places determined by uncontrollable factors. Tropical fig trees exhibit random distribution as well because of wind pollination. In addition to tropical fig trees and dandelion seeds, oyster larvae can travel hundreds of kilometers powered by sea currents, which causes random distribution when the larvae land in random places. Although random is thought to be unpredictable, it is the only dispersion that has a mathematical equation to represent it. This is due to the individualistic characteristics of random dispersion based on the idea that every species has equal opportunity and access to resources.

Species distribution can now be potentially predicted based on pattern of biodiversity at spatial scales. A general hierarchical model can integrate disturbance, dispersal and population dynamics. Based on factors of dispersal, disturbance, resources limiting climate, and other species distribution, predictions of species distribution can create a bioclimate range, or bioclimate envelope. The envelope can range from a local to a global scale or density independence to density dependence. The hierarchical model takes into consideration of requirements and impacts or resources as well as local extinctions in disturbance factors. Models can integrate the dispersal/migration model, the disturbance model, and abundance model. SDM's can be used to assess climate change impacts and conservation management issues. Species distribution models include presence/absence models, the dispersal/ migration models, disturbance models, and abundance models. A prevalent way of creating predicted distribution maps for different species is to reclassify a land cover layer depending on whether or not the specie in question would be predicted to habit each cover type. This simple SDM is often modified through the use of range data or ancillary information- such as elevation or water distance.

Recent studies have indicated that the grid size used can have an effect on the output of these species distribution models. The standard 50x50 km grid size can select up to 2.89 times more area than when modeled with a 1x1 km grid for the same specie. This has several effects on the species conservation planning under climate change predictions (global climate models- which are frequently used in the creation of species distribution models- usually consists of 50–100 km size grids) which could lead to over-prediction of future ranges in species distribution modeling. This can result in the misidentification of protected areas intended for a species future habitat.

The distribution of species into clumped, uniform, or random depends on different abiotic and biotic factors. Any non-living chemical or physical factor in the environment is considered an abiotic factor. There are three main types of abiotic factors: climatic factors consist of sunlight, atmosphere, humidity, temperature, and salinity; edaphic factors are abiotic factors regarding soil, such as the coarseness of soil, local geology, soil pH, and aeration; and social factors include land use and water availability. An example of the effects of abiotic factors on species distribution can be seen in drier areas, where most individuals of a species will gather around water sources, forming a clumped distribution.

Biotic factors, such as predation, disease, and competition for resources such as food, water, and mates, can also affect how a species is distributed. A biotic factor is any behavior of an organism that affects another organism, such as a predator consuming its prey. For example, biotic factors in a quail's environment would include their prey (insects and seeds), competition from other quail, and their predators, such as the coyote. An advantage of a herd, community, or other clumped distribution allows a population to detect predators earlier, at a greater distance, and potentially mount an effective defense. Due to limited resources, populations may be evenly distributed to minimize competition, as is found in forests, where competition for sunlight produces an even distribution of trees.

There are various ways to determine the distribution pattern of species. The Clark-Evans nearest neighbor method can be used to determine if a distribution is clumped, uniform or random. To utilize the Clark-Evans nearest neighbor method, researchers examine a population of a single species. The distance of an individual to its nearest neighbor is recorded for each individual in the sample. For two individuals that are each other's nearest neighbor, the distance is recorded twice, once for each individual. To receive accurate results, it is suggested that the number of distance measurements is at least 50. The average distance between nearest neighbors is compared to the expected distance in the case of random distribution to give the ratio:

$$\frac{mean\ distance}{\frac{1}{2}\sqrt{\text{density}}}$$

If this ratio (R) is equal to 1, then the population is randomly dispersed. If R is significantly greater than 1, the population is evenly dispersed. Lastly, if R is significantly less than 1, the population is clumped. Statistical tests (such as t-test, chi squared, etc.) can then be used to determine whether R is significantly different from 1.

The Variance/Mean ratio method focuses mainly on determining whether a species fits a randomly spaced distribution, but can also be used as evidence for either an even or clumped distribution. To utilize the Variance/Mean ratio method, data is collected from several random samples of a given population. In this analysis, it is imperative that data from at least 50 sample plots is considered. The number of individuals present in each sample is compared to the expected counts in the case of random distribution. The expected distribution can be found using Poisson distribution. If the variance/mean ratio is equal to 1, the population is found to be randomly distributed. If it is significantly greater than 1, the population is found to be clumped distribution. Finally, if the ratio is significantly less than 1, the population is found to be evenly distributed. Typical statistical tests used to find the significance of the variance/mean ratio include Student's t-test and chi squared.

However, many researchers believe that species distribution models based on statistical analysis, without including ecological models and theories, are too incomplete for prediction. Instead of conclusions based on presence-absence data, probabilities that convey the likelihood a species will occupy a given area are more preferred because these models include an estimate of confidence in the likelihood of the species being present/absent. Additionally, they are also more valuable than data collected based on simple presence or absence because models based on probability allow the formation of spatial maps that indicates how likely a species is to be found in a particular area. Similar areas can then be compared to see how likely it is that a species will occur there also; this leads to a relationship between habitat suitability and species occurrence.

In considering the **distribution** of an insect, and particularly that of an introduced form, it is usually customary to include within the area of distribution every place at which a single specimen has ever been collected. The distribution so defined may be termed the systematic or absolute distribution.

The success is due to 2 outstanding features in insect:

1. Tendency to increase
2. Subsequent dispersal

Of the various factors responsible for insect distribution, the major are climate (as insects are *poikilothermous*), availability of food and evolutionary history of the place.

In nature three groups of insects are found:

1. Tropical insects with a death point of 10-15°C
2. Temperate insects with a death point of 0°C
3. Boreal(Artic) with a death point between –20°C to –30°C

Insects are also called as:

1. Eurythermal-insects having wide range of temperature tolerance.
2. Stenothermal-narrow range of temperature tolerance.

Within the absolute distribution there is a much smaller region in which the insect causes damage to crops, and becomes of economic. This inner area, within which all outbreaks have occurred, may be called the area of economic distribution.

Accordingly, three zones of insect abundance have been recognized:

(a) **Zone of natural abundance** – the climate is congenial for the insects to undergo optimal growth, such insects, if happen to become pests, are regular pests.

(b) **Zone of occasional abundance** – the climatic conditions are favourable for the insects to develop and, insects that happen to attain pest status are occasional pests.

(c) **Zone of possible abundance** – a zone to which the adult insects spread from the two other zones; they survive here for a while and rarely attain pest status.

Insects are found around the world, from the tropical rainforests to the Antartica. Throughout history, insects have evolved, including mountain and ocean groups, retaining characteristics along the way. The five zoogeographical regions are: Northern hemisphere, Ethiopian, East-Asian, Australian and New tropical.

Diversity of Insects

Indeed, insects are found in virtually every terrestrial and fresh-water environment on the face of the earth. They live everywhere from mountaintops in the Himalayas to tide pools at the seashore. Dozens of eyeless species spend their entire lives in the total darkness of underground caverns. At least 40 species live on the polar ice caps of Antarctica; others exist on the surface of glaciers, or near the edge of melting snowfields. A few species can survive in hot springs where the water temperature is 50°C; others inhabit deserts where water is scarce and daytime temperatures exceed 60°C. Insects live both above and below the surface of rivers, streams, lakes, and ponds. One species of water strider makes its home among floating seaweeds in the Sargasso Sea, nearly 2400 kilometers (1500 miles) from the nearest land mass. The petroleum fly (*Helaeomyia petrolei*) develops in pools of crude oil in the oil fields. A related species, the brine fly (*Ephydra cinera*), thrives in the high salinity of salt lakes. In fact, the only place on earth where insects are **not** plentiful is in the ocean depths. Most marine environments are filled by another successful group of arthropods, the Crustacea.

Zoogeographical Regions: Wallace (200) established six zoogeographical regions: **Palearctic, Ethiopian, Oriental, Australian, Nearctic and Neotropical.** These are not of equal distinction, and may be arranged in groups. Palearctic and Nearctic, the north temperate regions, are very closely related, and are often treated as subgroups of the Holarctic Region. Though there is invasion of the Nearctic from the Neotropical, and Palearctic from Oriental, the Palearctic and Nearctic have more in common, showing that climate and history are more important than present geographical contiguity. The Ethiopian is most clearly related to the Oriental Region, though the former has some exchange with the Mediterranean Sub-region of the Palearctic. Ethiopian and Oriental are sometimes combined as Paleotropical, which to some also includes the Australian Region.

Nearctic, Palearctic, Ethiopian and Oriental are sometimes grouped together as Aretogaea, while Neo-tropical is segregated as Neogaea and Australian as Notogaea. Two large islands with associated smaller islands having distinctive faunae, Madagascar and New Guinea, do not fit perfectly into the above systems. Another classification of regions is by type of barriers; the climate-limited regions being the Palearctic and Nearctic, the barrier-limited regions being the Neo-tropical, Australian and Malagasy, and the main regions (which are in part barrier, and climate-limited) being Ethiopian and Oriental.

To some workers, the zoogeographical regions are purely arbitrary divisions, and some feel that the drawing of outlines of common or average ranges of groups of species has no real significance because factors controlling distribution of different groups may be quite unrelated.

Diversity studies in Ecology

Alpha diversity (α-diversity) is the biodiversity within a particular area, community or ecosystem, and is usually expressed as the Species richness of the area. This can be measured by counting the number of taxa (distinct groups of organisms) within the ecosystem (eg. families, genera, and species). However, such estimates of species richness are strongly influenced by sample size, so a number of statistical techniques can be used to correct for sample size to get comparable values.

Simpson's Index

$$D = 1 - \frac{\sum_{i=1}^{S} n_i(n_i - 1)}{N(N-1)}$$

where,

- ☆ S is the number of species
- ☆ N is the total percentage cover or total number of organisms
- ☆ n_i is the percentage cover of a species or number of organisms of species i

Shannon Index

$$H' = -\sum_{i=1}^{S} p_i \ln p_i$$

where,

- ☆ S is the number of species. Also called species richness
- ☆ p_i is the relative abundance of each species, calculated as the proportion of individuals of a given species to the total number of individuals in the community: n_i/N
- ☆ n_i is the number of individuals in each species; the abundance of each species
- ☆ N is the total number of all individuals

Beta diversity (β-diversity) is a measure of biodiversity which works by comparing the species diversity between ecosystems or along environmental gradients. This involves comparing the number of taxa that are unique to each of the ecosystems.

It is the rate of change in species composition across habitats or among communities. It gives a quantitative measure of diversity of communities that experience changing environments.

At its simplest, beta diversity is the total number of species that are unique between communities. This can be represented by the following equation:

$$\beta = (S_1 - c) + (S_2 - c)$$

where,

S_1= the total number of species recorded in the first community, S_2= the total number of species recorded in the second community, and c= the number of species common to both communities.

Sørensen's Similarity Index

$$\beta = \frac{2c}{S_1 + S_2}$$

where,

S_1= the total number of species recorded in the first community, S_2= the total number of species recorded in the second community, and c= the number of species common to both communities. The Sørensen index is a very simple measure of beta diversity, ranging from a value of 0 where there is no species overlap between the communities, to a value of 1 when exactly the same species are found in both communities.

Whittaker's Measure

$$\beta = \frac{S}{\bar{\alpha}} \quad or \quad \beta = \frac{S}{\bar{\alpha}} - 1$$

where,

S = the total number of species recorded in both communities,

$\bar{\alpha}$ = Average number of species found within the communities.

Gamma diversity (γ-diversity) is a measure of biodiversity. It refers to the total species richness over a large area or region. It is the product of α diversity of component ecosystems and the β diversity between component ecosystems.

According to Whittaker (1972), gamma diversity is the richness in species of a range of habitats in a geographic area (*e.g.*, a landscape, an island) and it is consequent on the alpha diversity of the individual communities and the range of differentiation or beta diversity among them. Like alpha diversity, it is a quality which simply has magnitude, not direction and can be represented by a single number (a scalar).

Gamma diversity can be expressed in terms of the species richness of component communities as follows:

$$\gamma = S_1 + S_2 - c$$

Where, S_1= the total number of species recorded in the first community, S_2= the total number of species recorded in the second community, and c= the number of species common to both communities.

The internal relationship between alpha, beta and gamma diversity can be represented as

$$\beta = \gamma/$$

Species richness is the number of different species in a given area. It is represented in equation form as *S*.

Species richness is the fundamental unit in which to assess the homogeneity of an environment. Typically, species richness is used in conservation studies to determine the sensitivity of ecosystems and their resident species. The actual number of species calculated alone is largely an arbitrary number. These studies, therefore, often develop a rubric or measure for valuing the species richness number(s) or adopt one from previous studies on similar ecosystems.

Estimates and Causal Factors

There is a strong inverse correlation in many groups between species richness and latitude: the farther from the equator, the fewer species can be found, even when compensating for the reduced surface area in higher latitudes due to the spherical geometry of the earth. Equally, as altitude increases, species richness decreases, indicating an effect of area, available energy, isolation and/or zonation (intermediate elevations can receive species from higher and lower).

Latitude (Latitudinal Gradient)

- Species richness increases from high latitudes to low latitudes.
- The peak of the species richness is not at Equator. It is deduced that the peak is between 20-30°N. The tropics fall within this range which is about 24.3 degrees north and south and this attests to the fact that species richness and biodiversity is highest here.
- The gradient of species richness is asymmetrical about the equator. The level of species richness increase rapidly from the north region but decrease slowly from the equator to southern region.

Area Effect

The latitudinal gradients of the species richness may result from the effect of area. The area at lower latitudes is larger than that at higher latitudes, leading to higher species richness at lower latitudes than normal.

Productivity

The latitudinal gradients of species richness may be result from the energy available to the ecosystems. At lower latitudes, there are higher amounts of energy available because of more solar radiation, more resources (for example, minerals and water); as a result, even higher levels of species richness can be allowed at lower latitudes. However, there have been relevant studies showing that species richness and primary productivity are actually negatively correlated.

The Millennium Ecosystem Assessment, an international ecological effort initiated by the United Nations, states:

"In most ecosystems, changes in the number of species are the consequences of changes in major abiotic and disturbance factors, so that the ecosystem effects of species richness (number of species) per se are expected to be both comparatively small and very difficult to isolate. For example, variation in primary productivity depends strongly on temperature and precipitation at the global scale and on soil resources and disturbance regime at the region-to-landscape-to-local scales. Factors that increase productivity, such as nutrient addition, often lead to lower species richness because more productive species outcompete less productive ones. In nature, therefore, high species diversity and high productivity are often not positively correlated."

Area

The species-area relationship is commonly approximated as following equation: $S = cA^z$ or $log(S) = log(c) + zlog(A)$ where S is the number of species, reflecting the species richness (sometimes also called species diversity), A is the area given in hectares, and c and z are constants. c is the species richness factor, usually between 20 and 2000; z is the species accumulation factor, usually between 0.2 and 0.5. This equation was first described by Arrhenius in 1921 and explains the variation of species richness among different areas.

Sampling

Measurements of simple species richness are related to area size, but are also affected by sampling intensity. When relatively small numbers of samples are taken from a given area, the numbers of species identified will be lower than if greater numbers of samples are collected more, the greater the number of species will be found. Thus in some cases, measurements of species richness can be considered a statistical artifact.

Habitat Diversity

It is possible that larger area contain more habitats as it is said that larger area is more topographically and environmentally diverse. Therefore, there are more opportunities for more species to set up their populations due to higher habitat diversity.

Relationship between Endemism and Species Richness

The levels of endemism and that of species richness are frequently positively correlated. However, on some oceanic islands, there are high levels of endemism but the levels of species richness are quite low.

Other Methods for Measuring Biodiversity

Adjusting the Species Richness

The most common formula for working out Species Diversity is the Simpson's diversity index, which uses the following formula:

$$D = 1 - \frac{\sum_i n_i(n_i - 1)}{N(N-1)}$$

where,

$$N = \Sigma_i n_i$$

- ☆ D = diversity index
- ☆ n_i = number of individuals of species i
- ☆ N = total number of organisms of all species found, *i.e.*

As predicted by the intermediate hypothesis, a high D value suggests a site that has undergone intermediate disturbance, while a low D value could suggest either recent disturbance or protection from disturbance.

Usually used in studies of vegetation but can also be applied to animals.

In order to account for the probability of missing some of the actual total number of species present in any count based on a sample population, the Jackknife estimate may be employed:

$$S = n + \left(\frac{n-1}{n}\right)^k$$

where,

- ☆ S = species richness
- ☆ n = total number of species present in sample population
- ☆ k = number of "unique" species (of which only one organism was found in sample population)

Similarly the equation may also be noted as:

$$S = E + k\left(\frac{n-1}{n}\right)$$

where,

- ☆ E = the summation of number of species in each sample
- ☆ k = number of rare/unique species
- ☆ n = number of sample

As well, when looking at local diversity the appropriate formula to use is:

$$S = c(A)^z$$

where,

- ☆ c = a specific number for each taxa
- ☆ A = the area of study
- ☆ z = the slope perimeter

Other measures of biodiversity may also take into account the rarity of the taxa, and the amount of evolutionary novelty they embody.

Weakness

Species richness fails to take into consideration species evenness. Other measures of biodiversity, such as the Simpson index, the Shannon index, and the fundamental biodiversity parameter θ of the unified neutral theory of biodiversity take species evenness into consideration.

Chapter 4

Basic Principles of Abiotic Factors and their Generalized Action on Insects

a) Temperature

This is the most important physical factor which determines the duration of the various stages in the insect life cycle and consequently the number of generations during any period of time. It acts on insects in two fold manner

1. By acting directly on the survival and development which determine the abundance of a pest
2. Indirectly through food and other environmental factors such as moisture, rainfall, wind etc.

Depending on the maintenance of body temperature, animal kingdom is divided into:

1) Warm Blooded Animals (Homeothermic)

These animals maintain a constant body temperature within certain narrow limits irrespective of the temperature variations in the external environment. These are also called as 'Endothermic animals' because they rely on internal source of heat to compensate the lost heat to cooler surroundings. *e.g.* **Mammals**

2) Cold Blooded Animals (Poikilothermic)

These animals are not capable of maintaining constant body temperature .They do not have internal mechanism of temperature regulation and therefore their body temperature varies with that of the surroundings. These are also called as 'Ectothermic animals' as they depend upon the environment than the metabolic heat to raise their body temperature. *e.g.* **Insects**

3) Socio-homeothermic Animals

These organisms maintain their body temperature slightly above the atmospheric temperature and are able to air condition their nests. They maintain their own temperature inside their colony irrespective of the temperature outside. *e.g.* **Honey bees**

Temperature regulates the development, fecundity, feeding, movement and dispersal and distribution of insects.

Development: In general insects grow more rapidly in warm weather than in cold weather. A given species of insects is active within certain limits of temperature. In general the optimum temperature for the normal development of insects is 10 to 35°C and is known as **'Zone Of Optimum or Normal Development'.**

Depending on the development of insects at different temperature levels, the temperature is divided into different zones as follows:

60°C - Maximum Fatal Temperature

50°C - Zone of Fatal High Temperature

35°C - Zone of Inactivity due to High Temperature

Maximum Effective Temperature (Threshold of Development), Upper vital limit Zone of Optimum/Normal/Effective Temperature

10°C - Minimum Effective Temperature (Threshold of Development), Developmental zero Zone of Inactivity due to Low Temperature

-5°C - Zone of Fatal Low Temperature

–14°C - Minimum Fatal Temperature

1. Zone of Effective Temperature (10 to 35°C)

In which some development takes place, the limits of which are known as

a) **Minimum effective temperature or threshold of development (10 to -5°C):** at which on descending scale development ceases and on ascending scale the development starts. The growth of **poikilothermic** animals is arrested at 0°C and this temperature is called as **'Developmental zero'**

b) **Maximum effective temperature (35 to 50°C):** The upper vital limit at which on ascending scale the development ceases and on descending scale the development begins.

2. Temperature Zones of Inactivity

The temperature immediately above and below the zone of effective temperature are the **'Zones of inactivity'.** In these zones the insect is alive but there will not be any development and they can recover if removed to favourable temperature.

3. Fatal Zones of Temperature

Beyond the zones of inactivity are the **'Fatal high (50 to 60°C)'** and **'Fatal low (-5 to -14°C)'** temperature zones, indicating **'Minimum fatal (-14°C)'** and **'Maximum fatal (60°C)'**. Death at fatal high temperature is due to loss of water, coagulation of proteins, high metabolic rate, accumulation of toxins and affected enzymatic activity. Death at fatal low temperature is due to low enzymatic activity, low metabolic rate and freezing of body fluids. Some insects do not freeze but survive even under 0°C as their body fluids contain polyols like glycerols. If an insect is given a choice to move along a temperature gradient it prefers a narrow limit of temperature known as the **'Temperature preferendum' or 'Preferred temperature'.**

Thermal Constant

The total heat energy required to complete a certain stage of development in the life cycle or in the completion of a physiological process of a species is constant and is termed as thermal constant and measured in Day - Degrees. Under unfavourable seasonal temperature the insects suspend their activities. These are of two types

1) Hibernation

A period of suspended activity in individuals occurring during seasonal low temperature

2) Aestivation

A period of suspended activity of individual occurring during seasonal high temperature or in a dry weather.

Fecundity Insects

Fecundity will be maximum at moderately high temperatures and declines at upper and lower limits of favourable temperature. Aphids remain parthenogenetic under high temperature and many hours of sunshine while the opposite condition gives rise to oviparous forms.

Distribution

Tropical and subtropical conditions as in India are ideal for the distribution and establishment of insects. Mediterranean fruit fly Ceratitis caiptata Wiedemanncould not establish in England and North Europe since its immature stages cannot stand below 10°C. Mosquitoes are more abundant at 70 to 80°F but are rare at 112 to 113°F. Pink boll worm of cotton *Pectinophora gossypiella* (Saunders) is serious in Punjab where the temperatures is within 95.5°F in August and September and not present in West Pakistan due to high temperatures at that period (99° F).

Dispersal and Movement

Insects try to move away from unfavorable temperatures. The rice weevil *Sitophilus oryzae* (Linnaeus) is found in the upper layers of bins irrespective of whether the initial infestation started at the depth of the bin or at surface due to rise in temperatures *i.e.* when the temperature reaches 32°C, the adults migrate to cooler upper layer. Mass flight of desert locust *Schistocerca gregaria* (Forskal) or migration starts at 17 to 22°C and they do not migrate when temperature is in between 14 to 16°C. Adaptations to temperature: At high temperature, locusts expose minimum body surfaces to sun's rays by lying parallel to them while they expose maximum body surface to sun's rays at low temperature lying at right angle to them.

b) Moisture

Insect body consists of 80 to 90 per cent water. Aquatic larvae contain about 98 per cent while insects which feed on dry food like Tribolium sp, Sitophilussp etc. Constitute about 50 per cent. Water is generally lost through spiracles and integument. Insects cannot afford to lose more water than they take and hence conserve water depending on the situation.

Adaptations to Conserve Moisture

1. **Body pigments:** Insects develop dark pigment in cooler areas which help to absorb more heat from sun for raising body temperature. This aids in getting rid of excessive moisture from the body. Light colour in desert insects helps to reflect sun's rays and save them from excessive evaporation.
2. **Integument:** Well developed integument and fused sclerites in beetles end weevils aid in conserving body moisture. Waxy coating of integument also saves from excessive evaporation.
3. **Winglessness:** Grasshoppers and crickets in arid regions have poorly developed wings and some are wingless by which the area of evaporation is reduced.
4. **Pilocity:** Dense hairs on the body prevent evaporation.
5. **Form of body:** Oval and compressed body of some desert beetles protects them from hot winds. Some desert insects have burrowing habit by which they go into deeper layers of soil when sufficient moisture is not available.
6. By reabsorption of water from products of excretion.
7. Some insect's links ***Amsacta*** spp. enter into aestivation when dry conditions prevail. The fall of water content of body below a certain minimum proves disastrous to insects and if it is considerably above the normal (in very wet places) harmful effects like disease outbreaks are noticed in insects.

Humidity

Unlike in temperature, there are no definite ranges of favourable humidity to all insects. Different species and their different immature stages have their own range. Humidity affects the speed of development, fecundity, colour etc. If water content of the body is high, dry air accelerates the development. Locusts sexually mature quicker and the number of eggs laid, are more at 70 per cent R.H. Rhinooceros beetle develops dark chitin in moist air and light chitin in dry air. Survival is indirectly affected by extremely high humidity conditions that favours the spread of diseases in insects.

c) Rainfall

Relative humidity is dependent on rainfall. The total amount of rainfall distribution in time influences the abundance of insects in an area. More than 12.5 cm rain during May-June results in increase in soil moisture which is not favorable to the cutworms and hence forced to come out of the soil and fall a ready prey to their parasites and predators. On the other hand if the rainfall is less than 10 to 12.5 cm during summer, cutworms remain protected in soil and there is outbreak of the pest in next season. Hence, the outbreak of pest can be forecasted, if the number of wet days (0.8 cm) during May-July is noted. If there are less than 10 wet days there will be an increase of cutworms in the following year. If there are more than 10 wet days there will be a decrease. Desert locust does not lay eggs and even if laid does not hatch unless soil has sufficient moisture. Rainfall also plays an important role in movement of swarms of desert locust. Saturated condition of moisture is injurious for the development of spotted boll worm ***Earias fabia*** Stoll. Red pumpkin beetle ***Aulacophora foveicollis*** Lucas withholds eggs until it come across moist soil. Rain induces emergence of most of the insects from soil. *e.g.* Ants, termites, red hairy caterpillar, root grub beetles etc., emerge out from the soil after the receipt of rains.

d) Light

Sunlight is the greatest single source of energy for all most all biological systems. Light as an ecological factor has been defined as all shorter wavelengths of radiant energy up to and including the visible spectrum which is measurable. Wave lengths of visible parts of spectrum range from 4000 (Violet) to 7600 (Red) Angstroms. Light is a non lethal factor. It helps in orientation or rhythematic behaviuour of insects, bioluminescence, period of occurrence and inactivity. The different properties of light that influence organisms are illumination, photoperiod, wave length of light rays, their direction and degree of polarization.

Visible and ultra violet light influences the following:

1. **Growth, moulting and fecundity**: silkworms develop faster in light than in darkness. Grubs of ***Trogoderma*** also develop more rapidly in light. Moths of spotted boll worm of cotton and red hairy caterpillar lay most of their eggs during periods of darkness. The bean weevil lays more eggs in total darkness than in light.
2. **Other activities**: In honey bees there is a correlation between hours of sunshine and their activity. Orientation of animals through directed movements by light is called **phototaxis** which also depends on temperature, moisture, food and age. Green leafhopper, *Nephotettix* spp. is attracted to light on hot and humid evenings but is indifferent to it during dry weather. Chafer beetles, many moths pass the day in concealment. Cockroaches hide during day time. Dusk is most usual time for flight and copulation of moths, for emergence of winged white ants etc.

   Based on daily activity cycle, insects or animals are categorized as:

   - ✰ **Diurnal:** Insects which are active during daylight hours
   - ✰ **Nocturnal**: Insects which are active at night

- ☆ **Crepuscular:** Insects which are active at dusk
- ☆ **Photoperiodism:** The number of hours of light in a day length (24 hours) is termed as photoperiod and the response of organisms to the photoperiod (length of the day) is known as **photoperiodism,** photoperiod induces diapause.

Insects in which diapause is induced by long day are known as **short day species.** *e.g.*: Mulberry silkworm, *Bombyx mori* (Linnaeus).

While the insects in which diapause in induced by short day lengths are known as **long day species.** *e.g.* Pink bollworm of cotton, *Pectinophora gossypiella.*

Photoperiod also known to control mode of reproduction, body form etc. In reduced photoperiod sexual forms (winged) are produced in aphids.

3. **Oviposition:** Light stimulates oviposition in mantids and inhibits in Periplaneta sp.
4. **Pigmentation:** In dark areas, pigmentation develops in insects *i.e.,* dark colour develops in dark areas.

Bioluminescence: Famous luminous insects are the glow-worms and fireflies. The enzyme luciferase in the presence of oxygen and adenosine triphosphate (ATP) promotes the oxidation of luciferin. This causes the production of light in insects. In most cases, females produce flash of light to attract males for mating.

Use of light as a factor in insect management: Many insects are either attracted or repelled to artificial light and this reaction is known as phototaxis. Grubs of *Trogoderma* sp. show negative reaction and are termed photonegative species. Most of the moths are attracted to light and are known as photopositive or phototropic. Based on this principle artificial light can be employed as a source for attracting insects and there after they can be trapped and destroyed and these devices are known as light traps.

e) Other Factors

1. **Atmospheric pressure:** It is generally of little importance. Locust shows great excitement and abnormal activity about half an hour before the occurrence of storm when the atmospheric pressure is low. Drosophila flies stop moving when put under vacuum.
2. **Wind and Air currents**: Most of the insects will not take flight when speed of wind exceeds the normal flight speed. Air currents, especially in the upper air being strong, carries many insects like aphids white flies, scales etc. to far-off places and is an important factor in dispersal. Air movement may also be directly responsible for death of insects. Severe wind coupled with heavy rains because mortality and moisture evaporation from body surface of insects.

f) Edaphic (Soil) Factors

Loamy soils allow digging and burrowing operation and are usually favourable for insects. Agrotis sps.live in soil of fairly light texture in which they move around freely in response to daily or seasonal temperature and moisture changes.

Chapter 5

Implications for Abundance and Distribution of Organisms including Insects: Law of Minimum, Law of Tolerance and Biocoenosis, Systems Approach to Ecology

The Law of Minimum

The metabolic process by which an apple, a tomato or a portion of cabbage is made into hair, or muscle or nerve, or some of the cells of the eyes, or into a hormone of some of the ductless glands are beyond our present comprehension, although a few steps in this process are supposed to be known. What we do know is that if the foods we eat are to be made into tissues and secretions, they must contain adequate amounts and due proportions of all the elements needed in the production of these things. Just as we cannot make concrete with cement alone or with sand alone, but must, if our concrete is to be good, have due proportions of both, so, if we are to have good structures in our bodies, we must have adequate amounts and proper proportions of all the food elements that go into the construction of these structures.

To make the best use of any food element, such, for example, as the proteins, certain other food substances and possibly a variety of them must be available in the diet in definite proportions. No food element is of itself of more importance in the body than another. It is only by the combination of all the necessary ones that the whole may be obtained.

Liebig laid down a **"law of the minimum"** in these words: **"The development of living beings is regulated by the supply of whichever element is least bountifully provided."**

The law of the minimum implies that the nutritive value of any food-mixture, however abundant most of the food elements there in may be, is limited by the minimum quantity of any essential element it may contain; unless, as it happens in certain food constituents, the lacking factor may be synthesized from some of the more abundant food elements, fat from sugar, for example.

Calcium cannot be utilized in producing enamel in the absence of phosphorus. If phosphorus is present in inadequate amounts, the enamel will be of poor quality. If iron is lacking, there will also be poor enamel. If any one of the vitamins is lacking poor enamel is the result.

The united "action" of minerals, proteins, vitamins, etc., in the production of tissues and fluids in the body is called "synergistic action." Since they do not act, but are used, it may be more correct to say that the body makes correlated use of them. At any rate, knowledge of the "synergistic action" of the food elements and vitamins is of vast importance and helps to make clear to us the **Law of the Minimum.**

Tooth enamel is made up largely of calcium and phosphorus with small quantities of iron. But the fusion of these substances into enamel requires the presence of vitamines A, C and D. At least half a dozen elements and factors are essential to the production of this tissue. In the absence of either of these factors the enamel of the teeth cannot be made. It is folly to feed much calcium in an effort to produce good teeth and to ignore the other essential elements of the teeth.

Many different tissues are being built and repaired in the body at all times. These tissues are constantly engaged in a wide variety of activities. This means that there is a continuous use of vitamins, minerals, amino acids, sugars and fatty acids in the body. Consuming an abundance of a single vitamin or of a single mineral is not sufficient to meet the needs of the many "synergistic actions" that are in constant process in the body. All of the vitamins and all of the minerals are required. If all the nutritive elements are present in adequate quantities and in due proportions, all the synergisms of all the minerals, vitamins, amino acids, sugars, etc., are at work piling up their benefits.

It is well to keep in mind that these "synergistic actions" are far more complex and broader than our description of them would indicate. The many functions of the body are also synergistic. The function of one part of the body is dependent upon the functions of all other parts. Feeding vitamin B in certain forms of nervous trouble with the aim of remedying the nervous impairment, and ignoring the correlated functions of the body that support nerve function, is doomed to failure. All of the "synergistic chemical actions" in all parts of the body must be adequately provided for before the actions of any part can be ideal. This is the thing that makes specific diets so unsatisfactory.

The work of potassium in promoting the formation of proteins, fats and glycogen, is not essentially unlike the work ascribed to vitamins. Zinc is thought to be connected with the action of the vitamins, which it seems to be able to replace to some extent, at least in the animal organism. Lithium is also thought to influence metabolism in a way not unlike that of the vitamins. The same is true of copper, nickel and of arsenic. It should be borne in mind that little is known about the offices of zinc, lithium, copper, nickel and arsenic in the body and that experimenters, carrying out their experiments, have wholly neglected these elements.

In plant life and growth the **Law of the Minimum** has long been known. If perfect growth is to be procured in plants, these must find a certain minimum of each of several elements in the soil in which they are grown. If only half the needed amount of potassium, for example, is present, then, regardless of how abundant all the other soil elements may be, their normal utilization is reduced one-half. The rate of growth of the plant and its ultimate development are correspondingly depressed.

Prof. Osterhout showed that seeds placed in distilled water grow better than when placed in water containing but one salt, and that each salt exerts a specific toxic effect according to its chemical nature. He found that one salt "counteracts" the effects of another.

An excess of nitrogen in the soil retards the formation of grain, roots and tubers and gives rise to sickly plants. Nitrogen excess in the soil of rose beds leads to the production of soft, sappy tissues, a luxuriant growth of leaves, and roses with little or no perfume.

Plant processes, from germination to ripening of their seed, are a drama of physiological minerals and gases. Without a sufficiency of lime or potash in the soil, acids, sugars and starches cannot be formed; without iron no chlorophyll and no albumen can be formed; without silica no fibre or plant skeleton, and the like. Each mineral plays a definite role in the growth of healthy plants.

The actual amount of potassium requisite for plant growth is very small compared with the needed amount of carbon or nitrogen. But if this relatively small quantity is not available the utilization of other constituents in tissue growth or repair is always deficient.

If one essential food element, which normally should compose one per cent of the food eaten, is present in only half this amount, then, the body will only be able to utilize the other elements, in tissue building, in the same reduced quantity.

The organic whole–the unit–is official to animal nutrition. Not the sum total of the minerals consumed, but their relative proportions, determines the nutritive value of any given dietary.

Shelford's Law of Tolerance

Shelford, Victor Earnest was an American Zoologist and animal ecologist whose pioneering studies of animal communities helped to establish ecology as a distinct discipline. His animal communities in temperate America (1913) were one of the first book to treat ecology as a separate science. Shelford was educated at the University of Chicago and spent most of his career at the University of Illinois. He was the first president of Ecological Society of America (1916). In 1933 Shelford began study at the

University of Illinois of the correlation between population changes of animal and environment changes. He made many reports on cyclic changes in animal population. With Fredric E Clements in 1939 published Bio ecology, in which he developed the concept of the biome predominant vegetation, with its animals inhabitants, that characterizes a large geographical area. His well known book The Ecology of North America (1963) summarised the major biomes, which include tundra, coniferous forest, deciduous forest, grassland, and desert.

The law of tolerance explains the ecological factors to which an organism responds as maximum and minimum effects and value below a critical minimum or a factor in excess of the critical maximum would exclude certain organism from an environmental areas.

All environmental factors have a tolerable limit and it may be critical minimum and critical maximum and range between them is called limit of tolerance. Therefore, value lying between these critical limits will be in between the limits of tolerance for an organism. If the limits of tolerance are exceeded for a particular species, it will lead to disappearance of the species from the area as long as such conditions exist.

According to Shelford law of tolerance the presence and success of an organism depends upon the completeness of a compex of condition s. Abundance or failure of organism can be controlled by the qualitative or quantitative deficiency or excess with respect to any one of several factors which may approach limits of tolerance for the organism

In case this limit is crossed on either side or reaches to the limits of tolerance then this is a zone of physiological stress. The other features of law of tolerance are as follows:

(1) An organism may have a wide range of tolerance for one specific environmental factor tha may have relatively low tolerant range for another factor

(2) Those organism which have wide ranges of tolerance for all environmental factors are, suppose to be most widely distributed.

(3) Sometimes, the environmental conditions are not optimum for a specific species in regard to one specific ecological factor then the limits of tolerance may be redused with respect to other ecological factors.

(4) In nature , it is not always possible for an organism to live at optimum range of particular physical factors. Therefore, in such cases some other factors has greater importance

(5) When environmental factors are likely to be limiting, this is the period of reproduction for an organism.

The range of tolerance and the optimum condition are different for every geographical race of the same species. Therefore, the law of tolerance and limits of tolerance help in understanding the biological distribution under natural conditions and also in the appearance of variations within species population.

Biocoenosis

A **biocoenosis** (alternatively, **biocoenose** or **biocenose**) termed by Karl Möbius in 1877, describes all the interacting organisms living together in a specific habitat (or biotope). **Biotic community**, biological community, and **ecological community** are more common synonyms of biocenosis, all of which represent the same concepts.

Three related descriptors within an ecosystem are:

- ✰ zoocoenosis for the faunal community,
- ✰ phytocoenosis for the floral community,
- ✰ Microbiocoenosis for the microbial community.http://en.wikipedia.org/wiki/Biocoenosis - cite_note-0

The extent or geographical area of a biocenose is limited only by the requirement of a more or less uniform species composition.

System Approach to Ecology

Systems ecology is an interdisciplinary field of ecology, taking a holistic approach to the study of ecological systems, especially ecosystems. Systems ecology can be seen as an application of general systems theory to ecology. Central to the systems ecology approach is the idea that an ecosystem is a complex system exhibiting emergent properties. Systems ecology focuses on interactions and transactions within and between biological and ecological systems, and is especially concerned with the way the functioning of ecosystems can be influenced by human interventions. It uses and extends concepts from thermodynamics and develops other macroscopic descriptions of complex systems.

Systems ecology seeks a holistic view of the interactions and transactions within and between biological and ecological systems. Systems ecologists realise that the function of any ecosystem can be influenced by human economics in fundamental ways. They have therefore taken an additional trans-disciplinary step by including economics in the consideration of ecological-economic systems.

In the words of R.L. Kitching:

- ✰ *Systems ecology can be defined as the approach to the study of ecology of organisms using the techniques and philosophy of systems analysis: that is, the methods and tools developed, largely in engineering, for studying, characterizing and making predictions about complex entities, that is, systems..*
- ✰ *In any study of an ecological system, an essential early procedure is to draw a diagram of the system of interest ... diagrams indicate the system's* ***boundaries*** *by a solid line. Within these boundaries, series of components are isolated which have been chosen to represent that portion of the world in which the systems analyst is interested ... If there are no connections across the systems' boundaries with the surrounding* ***systems environments****, the systems are described as* closed. *Ecological work, however, deals almost exclusively with* open *systems.*

As a mode of scientific enquiry, a central feature of Systems Ecology is the general application of the principles of energetics to all systems at any scale. Perhaps the most notable proponent of this view was Howard T. Odum - sometimes considered the father of ecosystems ecology. In this approach the principles of energetics constitute ecosystem principles. Reasoning by formal analogy from one system to another, enables the Systems Ecologist to see principles functioning in an analogous manner across system-scale boundaries. **H.T. Odum** commonly used the Energy Systems Language as a tool for making systems diagrams and flow charts.

The fourth of these principles, the principle of maximum power efficiency, takes central place in the analysis and synthesis of ecological systems. The fourth principle suggests that the most evolutionarily advantageous system function occurs when the environmental load matches the internal resistance of the system. The further the environmental load is from matching the internal resistance, the further the system is away from its sustainable steady state. Therefore the systems ecologist engages in a task of resistance and impedance matching in ecological engineering, just as the electronic engineer would do.

Chapter 6

Basic Concepts of Abundance, Model v/s Real World, Population Growth Basic Models, Exponential v/s Logistic Models, Discrete v/s Continuous Growth Model

What is a Population?

A population is a collection of interbreeding individuals of the same species that live together in a region. Population ecology is the study of population and how they change over time in relation to the environment including the environmental influences on population density and distribution, age structure and variations in population size. Population size can be constant, increasing or decreasing.

Various Factors that Affect Growth Pattern of a Population

- ☆ **Environmental factors**: Every organism requires an optimum condition for its growth and proliferation. Polar bears will not survive if they are exposed to desert conditions.
- ☆ **Food resources**: Availability of food and water is very important in development of an organism and hence to maintain its healthy body so as to reproduce efficiently.

- ☆ **Species interactions**: One species may depend on other species for survival. One species might be the food of the other or existence of one species might be mandatory for the other. Two species might share the same food and habitat which eventually leads to competition. In such cases population of one species might depend on that of the other.
- ☆ **Natural calamity**: An occurrence of a natural calamity will cause a steep decline in the number of organism or might even wipe off the species.
- ☆ **Human interventions**: Human activities such as poaching, deforestation, pollution caused by nearby human settlements can cause a drastic decrease in the number of species.

If there is situation, such that none of these factors are present to disturb the growth and proliferation of an organism, then the organism will keep on multiplying. Such a growth pattern of an organism is known as exponential growth. Nearly all populations will tend to grow exponentially as long as the resources are available. This is the simplest type of population growth.

Exponential v/s Logistic Models

Exponential model is associated with the name of Thomas Robert Malthus (1766-1834) who first realized that any species can potentially increase in numbers according to a geometric series. For example, if a species has non-overlapping populations (*e.g.*, annual plants), and each organism produces R offspring, then, population numbers N in generations t=0,1,2,... is equal to:

$$N_1 = N_0 \cdot R$$

$$N_t = N_0 \cdot R^t$$

When t is large, then this equation can be approximated by an exponential function:

$$N_t = N_0 \cdot \exp(r \cdot t) = N_0 \cdot e^{rt}$$

There are 3 possible model outcomes:

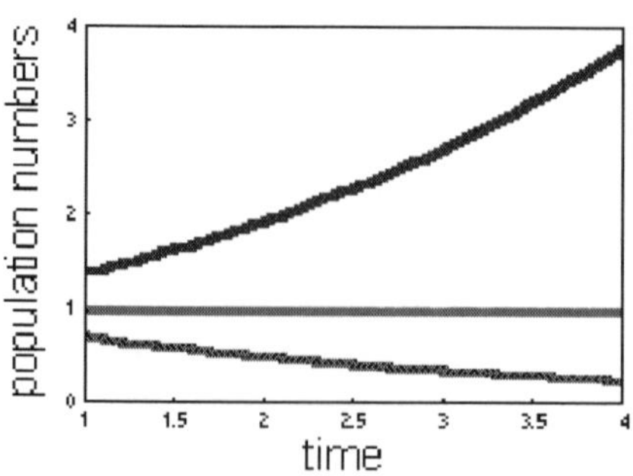

1. Population exponentially declines ($r < 0$)
2. Population exponentially increases ($r > 0$)
3. Population does not change ($r = 0$)

Parameter r is called:

- ☆ Malthusian parameter
- ☆ Intrinsic rate of increase
- ☆ Instantaneous rate of natural increase
- ☆ Population growth rate

"Instantaneous rate of natural increase" and "Population growth rate" are generic terms because they do not imply any relationship to population density. It is better to use the term "Intrinsic rate of increase" for parameter r in the logistic model rather than in the exponential model because in the logistic model, r equals to the population growth rate at very low density (no environmental resistance).

Assumptions of Exponential Model

1. Continuous reproduction (*e.g.*, no seasonality)
2. All organisms are identical (*e.g.*, no age structure)
3. Environment is constant in space and time (*e.g.*, resources are unlimited)

However, exponential model is robust; it gives reasonable precision even if these conditions do not met. Organisms may differ in their age, survival, and mortality. But the population consists of a large number of organisms, and thus their birth and death rates are averaged.

Parameter r in the exponential model can be interpreted as a difference between the birth (reproduction) rate and the death rate:

$$\frac{dN}{dt} = (b - m)N = rN$$

where,

b is the birth rate and

m is the death rate.

Birth rate is the number of offspring organisms produced per one existing organism in the population per unit time. Death rate is the probability of dying per one organism.

The rate of population growth (*r*) is equal to birth rate (*b*) minus death rate(*m*).

Applications of the Exponential Model

- ☆ microbiology (growth of bacteria),
- ☆ conservation biology (restoration of disturbed populations),

- insect rearing (prediction of yield),
- plant or insect quarantine (population growth of introduced species),
- Fishery (prediction of fish dynamics).

Population with Continuous and Descrete Growth

Exponential Growth

If a population has a constant birth rate through time and is never limited by food or disease, it has what is known as exponential growth. With exponential growth the birth rate alone controls how fast (or slow) the population grows.

Objectives

- To study the growth pattern of a population if there are no factors to limit its growth.
- To understand the various parameters of a population such as per capita rate of increase(r), per capita rate of birth (b) and per capita rate of death(d).
- To understand how these parameters affect the rate of growth of a population.

Exponential Population Growth Model

Exponential population growth occurs when a single species is not limited by other species (no predation, parasitism, competition), resources are not limited and environmental conditions are constant. In such conditions, population grows exponentially at constant percentage per time. Such a condition that permits exponential growth of a population is called an ecological vacuum. Ecological vacuum does not often occur in nature for a long period. In nature exponential population growth occurs commonly during a recovery of a population after a large scale disturbance (fire, epidemic. etc).

In an exponential population growth model, the change in population size may be determined by the following factors.

Change in population size during a fixed time interval=birth during time interval-deaths during time interval.

$$\frac{\Delta N}{\Delta t} = B - D$$

Birth Rate

Birth rate or natality rate is a measure of the extent to which a population replenishes itself through births.

$B = bN;$

b = Per capita birth rate

N = Population size

B = Number of birth

Death Rate

Death rate is the rate at which a population is losing individuals.

$D = mN;$

m = Per capita death rate

N = Population size

D = Number of death per year / number of individuals in a population

Exponential Population Growth Models may be Classified into:

1. Discrete population growth model.
2. Continuous population growth model.

Discrete Population Growth Model

An exponential population growth model may be defined as a discrete population growth model if the individuals of a population show:

- ☆ Discrete breeding season.
- ☆ Overlapping generations or Non overlapping generation.
- ☆ Semelparous life history or Iteroparous life history.

In **discrete breeding population** the species may breed only at a specific time usually at a particular time of the year. Breeding seasons introduce some delay in the regulative process. If a species lives for a number of years and produces relatively few young ones each year the delay time of one year due to discrete breeding season is likely to be short compared to the natural period of the species and so any oscillations caused by the delay will be convergent. But if the adults breeding in one season rarely or never survive to breed in the next has an important effect on their dynamics. In discrete growth we consider birth rate and death rate of the organisms.

In population of **overlapping generation** each generation lives for two periods like youth and old age (two period life versions). At any time period one generation of youth coexist with one generation of elderly. At the beginning of the next period the elderly die off, the youth themselves become elderly and a new generation of elderly is born. Thus there are two overlapping generation of people living at any one time. In **non overlapping generation** every period a new generation arises and old one dies off. Generation precedes and follows each other but they do not overlap at any point.

Semelparity and **iteroparity** refer to the reproductive strategy of an organism. A species is considered semelparous if it reproduces only a single time before it dies and Iteroparous if it can reproduce more than once in its lifetime.

In population with discrete population growth, the population growth depends on the R (geometric growth factor). Growth factor is the factor by which a quantity multiplies itself over time or fundamental net reproductive rate. Geometric growth factor is obtained from the difference in the number of birth per year and the number of death per year.

$$R = 1 + (B - D)$$

Imagine a situation where the government of India starts a sanctuary far from human settlements for protection of tigers (*Panthera tigris*). There is surplus food available in form of small mammals. Poaching is strictly prohibited. There is no other factor to disturb the intrinsic growth of the population. Mating can occur generally more common between November and April. The gestation period is 16 weeks.

If the population starts with 50 tigers and number of birth per year of the population from which the tigers were obtained was 0.45 and number of death per year was 0.1, then the R value will be 1+(0.45-0.1)=1.35 if so the population at the end of first year, second year, etc can be predicted. At the end of first year there would be:

$$N_1 = R.N_0 = (1.35)50 = 67.5$$

At the end of second year there would be:

$$N_2 = R.N_1 = R.R.N_0 = R^2 N_0 = 91.125$$

Thus at the end of t years the population will be:

$$N_t = R^t N_0$$

Continuous Population Growth

A population growth model may be defined as continuous population growth model if the individuals of a population show continuous breeding season. They never depend on parameters like season, food, climate for breeding. Here the population depends on instantaneous per capita rate of growth. It is best described in term of rate of population size change.

$$\frac{dN}{dt} = r * N$$

Here, r is the instantaneous per capita rate of growth. Here r can be calculated by the following equation:

$$r = \frac{\mathrm{N(t)} - \mathrm{N(0)}}{\mathrm{t}}$$

Consider the example of lion-tailed monkey. The Lion-tailed Macaque (Macaca silenus) is an Old World monkey that is endemic to the Western Ghats of South India. The Lion-tailed Macaque primarily eats indigenous fruits, leaves, buds, insects and small vertebrates in virgin forest. These animals are disappearing due to loss of habitat. Gestation is approximately six months. The young are nursed in the first year. Sexual maturity is reached at four years for females, six years for males. The life expectancy in the wild is approximately 20 years.

Let us suppose human settlement is withdrawn from a small area in the Western Ghats and the forest begin to grow rapidly, slowly the number of trees, insects and small vertebrates begin to increase. Let us say 102 lion-tailed monkeys find their way into this ideal place of growth and the growth of the lion-tailed monkeys are as follows:

Year	*Interval*	*Population in Numbers*
1972	0	102
1977	5	156
1982	10	202
1987	15	257
1992	20	302
1997	25	355

Then from the data: the rate of growth = (355-102)/25 = 10.12 = 0.1012.

Since the growth is between 0 and 1 make the data into decimal value by dividing by 100.

$$\frac{dN}{dt} = rN,$$

On Integrating

$$N_t = N_0 * e^r$$

Growth affects how common or rare a species is, which in turn affect the rate of population harvest. Like, how many fishes will be there after a period of trawling? How many trees can be cut from the population ensuring the same growth of the tree population? It is also important to protect the now endangered species. This can be easily done if we know growth rate of these species and protect them. This is especially useful in expanding the population. Thus though continuous or discrete mode of population growth is a very rare event it has its own importance in predicting the behavior of a population.

Calculate the growth using the equation and make it to decimals to run in the simulator;

$$r = \frac{\mathrm{N(t) - N(0)}}{\mathrm{t}}$$

Growth Rate
0.108
0.1
0.103
0.1
0.101

Balance of Life in Nature – Concepts of Carrying Capacity

Balance of Life

The population of an insect or any animal may be defined as the number of individuals of a particular species existing in a particular area at a time. The population never remains constant for long, but it tends to oscillate all the time about a theoretical optimum for the species. Balance of life in nature is the maintenance of a more or less fluctuating population density of an organism over a given period of time within certain definite upper and lower limits by the action of biotic and abiotic factors. Factors Contributing to Population Increase Any organism will multiply enormously if the environment is optimum. Different organisms multiply at different rates. Hence it is well known that every organism has an inherent capacity to survive, reproduce and multiply in numbers. The extent to which a species can multiply in a given period of time if no adverse factors interfere is called its 'Biotic potential' which is also known as 'Maximum reproductive power'. This concept was first introduced by R.N. Chapman in 1928. The biotic potential or innate capacity to increase depends on 1) Initial population: The more the initial population of an organism the more will be its progeny, 2) Fecundity: It is the average number of eggs laid by a female in its life. The more the fecundity the more will be the resultant population. 3) Sex ratio: It is the ratio of females to the total population and is represented by number of females/ Total number of males and females. Up to a limit the more the proportion of females, the more the multiplication capacity. 4) Number of generations in an unit time or a year: Obviously the greater the number of generations in a unit time the larger will be the resultant population.

Based on the above factors the biotic potential can then be represented by the formula,

B. P. = P (f s) n

where,

P = Initial population
f = Fecundity
s = Sex ratio
n = Number of generations in a unit time.

Some insects like whiteant queens and house flies lay large number of eggs while others lay very few eggs. Some insects reproduce very fast. Mustard aphid has over 40 generations a year. If all survive, a single pair of house flies may produce 191, 010, 000, 000, 000, 000, 000, flies from April to August which if spread over the entire earth form a layer about 14 meters deep. Similarly a progeny of a pair Drosophila flies produced in a year would cover the whole of Indian subcontinent and Myanmar with a solid cake of flies. Such is the biotic potential of insects when there is no interference of biotic and abiotic factors of the environment.

Factors Tending to Reduce Populations: However, in nature there are other powerful factors working against the biotic potential. These are

(1) Abiotic or climatic factors and (2) Biotic factors. These biotic and abiotic factors are known as the constituents of environmental resistance which always tend to destroy a considerable proportion of insect life. The proportion of the population which is normally eliminated as a result of environmental resistance is known as 'Normal Coefficient of destruction' which can be expressed by the formula,

$$Qn = 1 - (1/s)\, n/fn.$$

where,

Q = The coefficient of destruction,

s = The sex ratio when population is taken as 1

n = The number of generations in a unit interval of time and

f = Fecundity.

Balance of Life In nature there are two sets of tendencies namely the biotic potential tending to increase the population and the environmental resistance tending to reduce the population. As such there is a constant interaction between these two opposing forces and then maintains a dynamic equilibrium known as 'Balance of life'. It is evident from the above that in any case, the insects or other animals never attain the high density which they are potentially capable of doing which is because of environmental limiting factors like abiotic factors comprising mainly temperature and humidity which at too high or too low levels adversely affects insects. Natural disturbance like heavy rain, hail storms, snow, sand storms, dust storms, and very high wind velocity are adverse to insect life. Biotic factors *i.e.* limitation of food, competition for food and space and natural enemies act adversely depending on the density of population

Environmental Resistance

In nature there are two sets of tendencies namely the biotic potential tending to increase the population and the **environmental resistance** tending to reduce the population. As such there is a constant interaction between these two opposing forces and then maintains a dynamic equilibrium known as 'Balance of life'. It is evident from the above that in any case, the insects or other animals never attain the high density which they are potentially capable of doing which is because of environmental limiting factors like abiotic factors comprising mainly temperature and humidity which at too high or too low levels adversely affects insects. Natural disturbance like heavy rain, hail storms, snow, sand storms, dust storms, and very high wind velocity are adverse to insect life. Biotic factors *i.e.* limitation of food, competition for food and space and natural enemies act adversely depending on the density of population.

Chapter 7

Vital Statistics: Life Table and their Application in Insect Biology

Life Table

Life table concept was first formulated by Raymond Pearl (1924). Morris and Miller (1954) were the first to adopt the life table format for the study of natural insect populations. They studied in the spruce bud worm, *Choristoneura fumiferana*.

Life table is the systematic tabulation of births and deaths of an organism. It is summary statement on the life of a typical individual of population or a cohort of individuals.

A. Typical Life Table Contains

X= Age of cohort

d_x = Number dying at age interval

l_x = Number surviving at the beginning of age interval.

1000 q_x = Mortality rate (per thousand) at the beginning of age interval or d_x as a percentage of l_x.

e_x = Expectation of life or mean life time remaining to those attaining age interval.

The age interval x is into life stages *i.e.* egg, larval stages pupae and adult and additional column d_xF was added the factor responsible for d_x. This d_xF column takes into account all the mortality factors that could be quantified and the d_x and 1000 q_x column take into account actual values for each of these factors.

x	l_x	d_xF	d_x	1000 q_x
Age interval	No. alive at beginning of x	Factor responsible for d_x	No. dying during x	d_x as percentage of l_x

If the logarithm of number of a species is plotted against time or age a curve called survivorship curve is produced. The shape of the curve is an indication of the mortality distribution with age.

Three types of curves are recognized. They are

1. Type I curve: It shows a very low death rate of young and high mortality in old age. Ex: Higher vertebrates and Drosophila.
2. Type II curve: It is straight line survivorship curve. Ex: Birds, Hydra, Bacteria.
3. Type III curve: in this curve huge mortalities occure in early stages. Ex: Parasitic organisms.

Life Tables are of 2 types

1. Age dependent
2. Stage dependent

a. Age-dependent Life Tables

Age -dependent life tables shows organisms mortality (or survival) and reproduction rate (maternal frequency) as a fu ıction of age. in nature , mortality and reproduction rate may depend on numerous factors : temperatures, population density etc. when building a life table , the effect of these factors is averaged. Only age is considered as a factor that determines mortality and reproduction. Only females are considered in this life table.

b. Stage-dependent Life Tables

These are used mainly for insects and other terrestrial invertebrates.

Stage dependent life tables are builds in the cases when:

- ☆ The life cycle is partitioned into distinct stages (eggs, larva, pupae and adults in insect).
- ☆ Survival and reproduction depend more on organism stage rather than on calendar age.
- ☆ Age distribution at particular time does not matter (ex there is only one generation per year).

Advantages and Limitation of Life Tables

Advantages

- ☆ Life tables provide age specific data which are ideal for use in population management.
- ☆ It gives information about population trend in relation to environment factors.
- ☆ It helps in reducing the out breaks of insect pest by altering the factors.
- ☆ Useful for insects having one generation and little overlapping of developmental stages.

Limitations

- ☆ Sampling of different developmental stages of same insects at different locations is very difficult.
- ☆ It is not useful for insects having widely overlapping stages and generations

Survivorship Curves

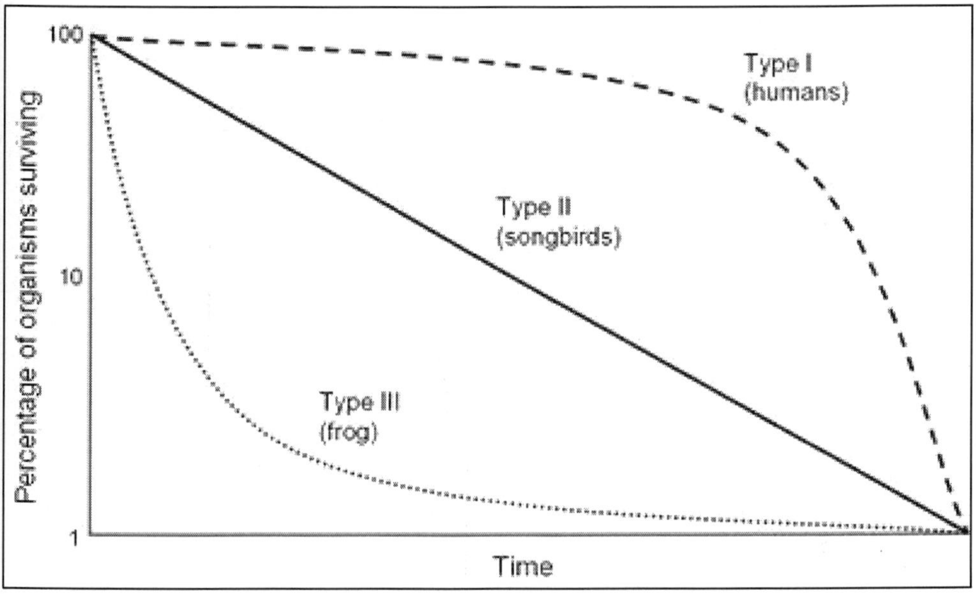

An illustration of the three idealized types of curve

A survivorship curve is a graph showing the number or proportion of individuals surviving to each age for a given species or group (*e.g.* males or females). Survivorship curves can be constructed for a given **cohort** (a group of individuals of roughly the same age) based on a life table.

There are three generalized types of survivorship curves:

- ☆ Type I survivorship curves are characterized by high age-specific survival probability in early and middle life, followed by a rapid decline in survival in later life. They are typical of species that produce few offspring but care for them well, including humans and many other large mammals.
- ☆ Type II curves are an intermediate between Types I and III, where roughly constant mortality rate/survival probability is experienced regardless of age. Some birds and some lizards follow this pattern.
- ☆ In Type III curves, the greatest mortality (lowest age-specific survival) is experienced early in life, with relatively low rates of death (high probability of survival) for those surviving this bottleneck. This type of curve is characteristic of species that produce a large number of offspring. This includes most marine invertebrates. For example, oysters produce millions

of eggs, but most larvae die from predation or other causes; those that survive long enough to produce a hard shell live relatively long.

The number or proportion of organisms surviving to any age is plotted on the y-axis, generally with a logarithmic scale starting with 1000 individuals, while their age, often as a proportion of maximum life span, is plotted on the x-axis.

In mathematical statistics, the survival function is one specific form of survivorship curve and plays a basic part in survival analysis.

Mortality Patterns and Survivorship Curves

Population growth patterns require an assumption that members of a population are all identical individuals. However, individuals are in their different stages of lifespan. In a given period of time, some are born and some die. See if you can figure out the data in this table:

A Life Table for a *Poa annua* (a grass) Cohort

Age (Month)	*Number Observed Alive*	*Proportion Surviving*	*Number Dying*	*Mortality Rate Per Capita*	*Avg. Number of Seeds/ Individuals*
0-3	843	1,000	121	0.143	0
3-6	722	857	195	0.271	300
6-9	527	625	211	0.400	620
9-12	316	375	172	0.544	430
12-15	144	171	95	0.626	210
15-18	54	64	39	0.722	60
18-21	15	17.8	12	0.800	30
21-24	3	3.6	3	1.000	10
24	0	–	–	–	–

A *cohort* of a population is the total number of new births of it at the same time. As you can see, the number of grass observed alive at the beginning is 843, and data are noted down every 3 months. The grass is gradually dying out, and the life stages at which they perish vary. From another perspective, let's look at the number that survives instead of focusing on the number dying in each period. After the first observational period, 722 survive the first 3 months. ***Survivorship* is *the probability of newborn individuals of a cohort surviving to particular ages*.** Plotting the number surviving against percent of life span, we get survivorship curves that show the number of individuals of a cohort still living over time.

For the sake of discussion, we will establish three types of representative survivorship curves hypothetically, displayed in the upper left graph of the following image. Curve I is characteristic of a population in which most individuals survive well pass the midpoint. On the contrary, Curve III typifies populations wherein most individuals die young. In the type II curve, survivorship decreases at a constant rate throughout the lifespan.

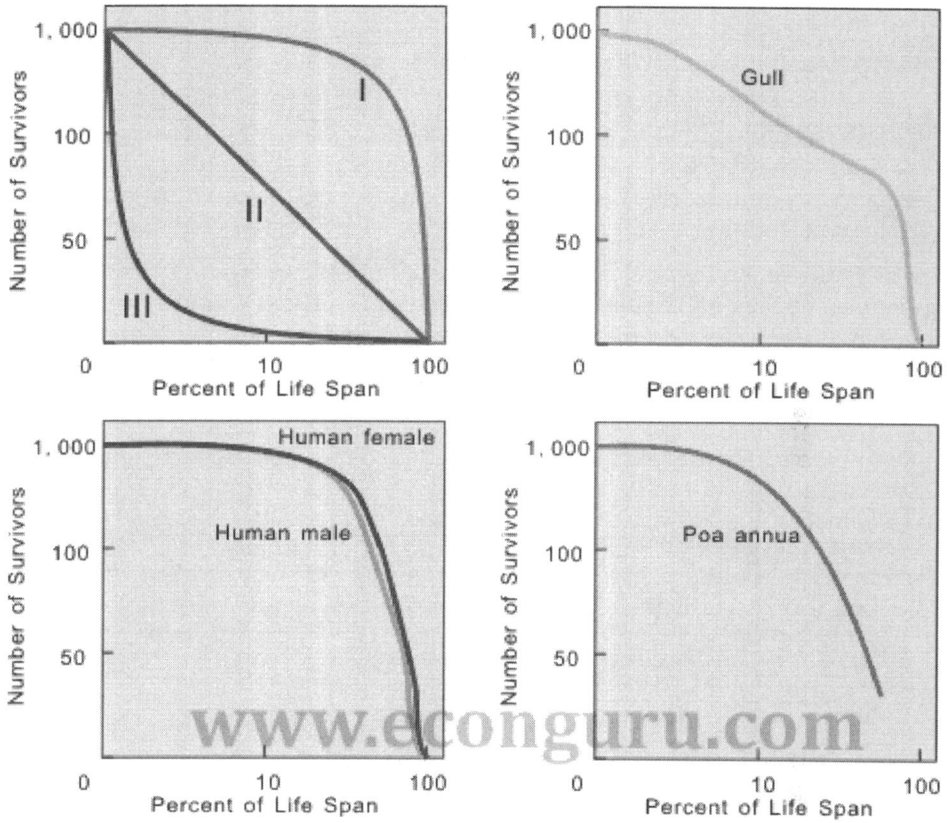

Survivorship Curves

As shown by the other three graphs, the survivorship curves of natural populations do not fit these three idealized curves congruously. However, similarities exist to help understanding. Survivorship curve for *gulls* fits the type II curve somewhat. The survivorship curves of *human males and females* differ slightly but both resemble the type I curve fairly well. The survivorship curve for *Poa annua* seems to be a combination of the type I and type II curves.

Survivorship curves denote mortality patterns of a certain population over a certain period of time. It may vary in abnormal conditions, but in most cases the pattern stay predictable.

Case Study of Insect Life Table

A **"life table"** is a kind of **bookkeeping system** that ecologists often use to keep track of stage-specific mortality in the populations they study. It is an especially useful approach in entomology where developmental stages are discrete and mortality rates may vary widely from one life stage to another. From a pest management standpoint, it is very useful to know when (and why) a pest population suffers high mortality – this is usually the time when it is most vulnerable. By managing the natural environment to maximize this vulnerability, pest populations can often be suppressed without any other control methods.

To create a life table, an ecologist follows the life history of many individuals in a population, keeping track of how many offspring each female produces, when each one dies, and what caused its death. After amassing data from different populations, different years, and different environmental conditions, the ecologist summarizes this data by calculating average mortality within each developmental stage.

For example, in a hypothetical insect population, an average female will lay 200 eggs before she dies. Half of these eggs (on average) will be consumed by predators, 90 per cent of the larvae will die from parasitisation, and three-fifths of the pupae will freeze to death in the winter. (These numbers are averages, but they are based on a large database of observations.)

The complete life table from the preceding data would be as follows:

Developmental Stage	Number Alive	Mortality Factor(s)	Number Dying	Percent Mortality
Egg	200	predation	100	50
Larva	100	parasitization	90	90
Pupa	10	freezing	6	60
Adult	4	(2 females and 2 males if sex ratio is 1:1)		

A life table can be created from the above data. Start with a **cohort**of 200 eggs (the progeny of Mrs. Average Female). This number represents the maximum **biotic potential** of the species (*i.e.* the greatest number of offspring that could be produced in one generation under ideal conditions). The **first line** of the life table lists the main cause(s) of death, the number dying, and the percent mortality during the **egg stage**. In this example, an average of only 100 individuals survive the egg stage and become larvae. The **second line** of the table lists the mortality experience of these 100 larvae: only 10 of them survive to become pupae (90 per cent mortality of the larvae). The **third line** of the table lists the mortality experience of the 10 pupae – three-fifths die of freezing. This leaves only 4 individuals alive in the **adult stage** to reproduce. If we assume a 1:1 sex ratio, then there are 2 males and 2 females to start the next generation.

If there is no mortality of these females, they will each lay an average of 200 eggs to start the next generation. Thus there are two females in the cohort to replace the one original female – this population is **doubling** in size each generation!!

In ecology, the symbol **"R"** (capital R) is known as the **replacement rate**. It is a way to measure the change in reproductive capacity from generation to generation. The value of **"R"** is simply the number of reproductive daughters that each female produces over her lifetime:

$$R = \frac{\text{Number of daughters}}{\text{Number of mothers}}$$

- ☆ If the value of "R" is less than 1, the population is decreasing – if this situation persists for any length of time the population becomes extinct.
- ☆ If the value of "R" is greater than 1, the population is increasing – if this situation persists for any length of time the population will grow beyond the environment's carrying capacity. (Uncontrolled population growth is usually a sign of a disturbed habitat, an introduced species, or some other type of human intervention.)
- ☆ If the value of "R" is equal to 1, the population is stable – most natural populations are very close to this value.

Practice Problem

A typical female of the bubble gum maggot (*Bubblicious blowhardi* Meyer) lays 250 eggs. On average, 32 of these eggs are infertile and 64 are killed by parasites. Of the survivors, 64 die as larvae due to habitat destruction (gum is cleared away by the janitorial staff) and 87 die as pupae because the gum gets too hard. Construct a life table for this species and calculate a value for "R", the replacement rate (assume a 1:1 sex ratio). Is this population increasing, decreasing, or remaining stable?

Life Table Studies of *Trilocha virescence* (Bombycidae: Lepidoptera) on *Ficus nitida*

Computing life tables is a fairly current approach used by entomologists to study the insects' population dynamics. With the recent emphasis on population dynamic studies of insects, the usefulness of life table in this area is becoming more recognized. A life table is a tabular device which describes for every particular age of interval. In ecological study, life table is an important analytical tool, which provides detailed information of population dynamics to generate simple but more informative statistics. It also gives a comprehensive description of the survivorship, development and expectation of life (Ali and Rizvi, 2007; Yzdani and Samih 2012; Gabre *et al.*, 2004). Life tables may be used to study inherent differences in the survivorship and reproductive strategies of populations under different ecological regimes (Afrane *et al.*, 2007). Life table, studies provide an opportunity to assess and evaluate the impact of specific mortality factors acting on insect population (Harcourt, 1969; Bellows *et al.*, 1992; Mohapatra, 2007). It can also give information about the changes which occur during the developmental stages and their relationship with different environmental factors (Atwal and Bain, 1974). Life tables constructed using laboratory data collected under controlled conditions and are useful revealing the maximal growth potential of a population (Gabre *et al.*, 2004). Life tables used can make quantitatively and qualitatively evaluation of various host plants (Ambegaonkar and Bilapate, 1981). Two kinds of life tables have been developed: (1). ecological life tables for natural populations of animals; and, (2) life tables for laboratory populations of animals. They may be further categorized as horizontal or age-specific life tables and vertical or time-specific life tables. The horizontal life table usually is made from data obtained by studying all or a portion of a population belonging to a single generation. Vertical life tables involve analysis of the age structure of a population (Southwood, 1966). Age specific life tables provide a concise summary of mortality and reproductive schedules and can help explain why certain species proliferate in particular environment (Afrane *et al.*, 2007). Trilocha virescence is a member of family Bombicidae and important pest of Ficus nitida, which is an ornamental shurb of medium size. T. viresencecan be considered as a voracies feeder of Ficus nitida leaves. Itslarvae in 4th and 5th stages are capable of eating more than one leaf per day and can cause severe defoliation of the plant in case of severe infestation. So the studies were conducted with the aim to find out survivorship, mortality, life expectancy and reproductive rate of Trilocha virescence onFicus nitida.

Trilocha virescence Culture

For conducting life table of *T. virescence* under ambient condition, its egg batches were collected from Ficus nitida and reared in Biological Control Laboratory, Department of Entomology, PMAS Arid Agriculture University, Rawalpindi. These eggs were kept in petri dishes provided with slightly moistened filter papers and allowed to hatch. From these batches 100 newly emerged larvae were selected and shifted in petri dishes with the help of camel hair brush. Each petri dish was provided Ficus nitida tender leaves as a larval food. All transferred larvae were checked daily and their molting to the next stage or mortality were recorded carefully. Subsequent stages like pupa and adult were also checked. After the emergence of adults, male and female were separated. Pairs were made, and each pair was placed in a separate petri dish. Adults were provided with special diet that includes water, methyl paraben, insect's vitamin and sugar. Small pieces of cotton soaked in the diet were placed in the petri dish until the adults completed their lives. Dead males were replaced with the new ones. Fecundity and survival rate of each female were recorded until their death.

Life Table Construction

Stage Specific Life Table

For calculating this life table, stage specific data regarding

Mortality and survival of different insect stages were taken from the age specific life table.

Following parameters were used to compute stage specific life table.

x = Age of the insects in days

lx = Number surviving at the beginning of each interval, out of 100

dx = Number dying during the age interval, out of 100

The information from above assumptions was used to compute different life table parameters as follows:

Apparent Mortality (100qx): The information regarding number of insects dying as a percent of number entering a particular stage can be obtained from the apparent mortality, which can be calculated as:

Apparent Mortality = [dx/lx] × 100

Survival Fraction (Sx): The information obtained from the apparent mortality can be used to calculate survival fraction of each stage by applying the formula:

Tx = lx + (lx+1) + (lx+2)...........+ lw

where,

lw = Last age interval

ex = Expectation of life or mean life for individuals of age x, which was calculated by using following equation

ex = Tx/lx

Cohort Life Table of *Trilocha virescence*

The larvae were observed daily to determine the dead and alive ones. Following parameter were used to construct the life table of *Trilocha virescence* on *Ficus nitida*.

x = Age of the insects in days

lx = Number surviving at the beginning of each interval, out of 100

dx = Number dying during the age interval, out of 100

100qx = Mortality rate at the age interval x and calculated by using formula

100qx = [dx/lx] × 100

To calculate ex, two other parameters like Lx and Tx were also computed

Lx = Number of individuals alive between age x and x+1

Lx = lx+1 (x+1)/2

Tx = The total number of individual of x age units beyond the age x

Tx = lx + (lx+1) + (lx+2)...........+ lw

where,

lw = Last age interval

ex = Expectation of life or mean life for individuals of age x, which was calculated by using following equation

ex = Tx/lx

Average daily mortality = 1/ex

Fx = eggs produced at each stage

mx = eggs produced per surviving individual at each stage

lxmx = eggs produced per original individual at each stage

Chapter 8

Population Dynamics – Factors Affecting Abundance – Environmental Factors, Dispersal and Migration, Seasonality in Insects, Classification and Mechanism of Achieving different Seasonality – Diapauses (Quiescence) – Aestivation, Hibernation

Population Dynamics

Population ecology is a major sub - field of ecology that deals the dynamics of species populations and how these populations interact with environment.

Population dynamics is the branch of life sciences that studies, short and long-term chances in the size and age composition of populations, and the biological and environmental processes influencing those changes.

Population dynamics deals with the way populations are affected by birth and death rates, and by immigration and emigration, and ageing populations or population decline.

Population is affected by 3 dynamic rate functions. They are:

1. Natality is birth rate, often recruitment, which means reaching a certain size or reproductive stage.
2. Population growth rate, which measures the growth of individuals in size and length.
3. Mortality, which includes harvest mortality.

If N1 is the number of individuals at time 1 then

$N_1 = N_0 + B - D + 1 - E$

Where N_0 is the number of individuals at time 0, B is the number of individuals born , D the number that died, 1 the number that immigrated, and E the number that emigrated between time 0 and time 1 .

- Intrinsic rate of increase

The rate at which a population increases in size if there are no density – dependent forces regulating the population is known as the *intrinsic rate of increase.*

$(dN/dt)(1/N) = r$

Where (dN/dt) is the rate of increase of population

N is the population size

R is the intrinsic rate of increase

An important concept in population ecology is the r/K selection theory.

r is the intrinsic rate of increase in population size. Density dependent

K is the carrying capacity of a population, density dependent

An r – selected species (aphids) is one that high rates of fecundity, low levels of parental investment in the young, and high rate of mortality before individuals reach maturity. Evolution favours productivity in r – selected species.

K- Selected species (such as humans) has low rates of fecundity, high levels of parental investment in the young and low rates of mortality as individuals mature. Evolution in K- selected species favours efficiency in the conversion of more recourses into fewer offsprings.

Factors Affecting Abundance – Environmental Factors

Environmental factors affect insect populations. A limiting factor exceeds the tolerance level for a species, *e.g.* low winter temperature in ticks. Determining a limiting factor usually requires lab studies as there are many factors at work in the field.

Important in the distribution of insects -daily conditions more important than averages.

A. Light

- May control vegetation and therefore insect populations
- Photoperiod -stimulus for diapause initiation
- Night v day used by insects whether nocturnal or diurnal

B. Temperature

Body temperature of insects governs rate of growth. Since insects have limited control over body temperature, outside temperature is important.

Most insect activity and reproduction occurs between 15 -35o C but each species usually has an optimum temperature.

Note that:

- Within a species different stages may have different rates of development.
- One stage may be more resistant than another stage *e.g.* over wintering stage of bronze orange bug.
- Rates of growth of insects and plants closely tied together *e.g.* silkworm development is tied to the appearance of mulberry leaves

C. Wind

- Indirect effect causing evaporation, humidity and so desiccation
- Dispersal *e.g.* aphids, moths grasshoppers

D. Moisture

Rain

- Directly by mechanical effect, disturbing aphids
- Indirectly by affecting soil and water supply for host plants

Humidity

Important for egg hatching activity and development.

E. Interaction between Temperature and Moisture

A = Region of most rapid development

B = Region of favourable development

C = Region of retarded development

D = Region of no development

It can be seen that development is adversely affected at both temperature extremes but development is favoured at higher relative humidities.

Where a particular species lives - forest, savannah etc or terrestrial aquatic etc

The type of soil in a habitat influences an insect's distribution and abundance and is easily disturbed by agriculture, *e.g.* Irrigation changes moisture and subsequently, the type of pest in a crop. Chemicals in soil affect plant growth and therefore the dependant insects.

Niche = physical location and function of a species at a given time. If two species occupy the same niche, they will compete and one will displace the other.

- Starvation -blowflies needing flesh may not find necessary corpses or wounds, silkworms hatching before bud opening of mulberry leaves.
- Decline in supply - cabbage aphids at end of seasonal crop or after harvest
- Dependence up food chain *e.g.* predators of above cannot find prey

- ☆ Lack of food at critical time - often dependent upon weather
  - Hover flies (syrphid) require spring pollen for maturation of ovaries. If plants not in flower –> few eggs.
  - Sorghum midges diapause over winter as mature larvae, emerging with summer rain as adults for only 2 days and need to find sorghum flowers in that time.

Effective storage

Potential food exists but is unavailable for consumption

- ☆ Accidental loss of food - lice, fleas fall off host; aphids, grasshoppers blown by wind
- ☆ Interference by other species - humans apply repellent to crops or livestock
- ☆ Insect behaviour. Tsetse flies don't feed on animals in the open. If host is not close by shade, they will not feed.
- ☆ Cannibalism - If codling moth larvae encounter another in an apple, one is eaten despite plenty of apple for both.
- ☆ Affects of insect feeding on hosts - If buffalo flies are in large numbers, their host will take action to avoid being bitten.

Food Quality

Nutrient deficiencies. Mite numbers, egg production and longevity are directly related to N_2 content in leaves.

Development time in mites is indirectly related to the N_2 content of leaves.

Plant variety - cultivated varieties supported greater populations of aphids than wild beans. [One female aphid placed on wild beans produced a population of 37 after 14 days. However, when placed on each of 19 bean varieties, a single female gave populations of between 286 - 1037]

Different foods produced different development and survival rates in moth (*Ephestia*) larvae and similarly for Thrips:

Note that although the addition of pollen, (high in protein), may or may affect survival, it greatly enhances egg prduction.

Different foods can produce different castes of insect. *e.g.* Exclusive feeding of royal jelly to bee larvae produces queens rather than workers.

How do Insects Overcome Food Problem?

- ☆ Dispersal
- ☆ Polyphagy - eat multiple species of predators or plants
- ☆ Stoage of food - social insects - ants, bees.

Other Organism

Intraspecific between individuals of the same species. This involves population density scarcity and crowding

At low density males and females may not meet *e.g.* sheep keds. These wingless flies move from sheep to sheep by sheep contact. Ked population will die out id there are less than one per sheep.

At high density there is competition between individuals. This leads to reduced reproductive rate, survival and longevity.

Interspecific competition -between individuals of other species. Involves niche overlap whereby the needs of 2 or more different species for a resource coincide. This leads to competitive displacement. *e.g.* Two species of flour beetle (*Tribolium confuse* and *T. cutaneum*) were placed in flour at 29 C and counted every 30 days for 3 years.

T. casteneum displace *T. confusum* under these conditions.

However, if the temperature is < 29°C, *T. confusum* dominate.

Natural Enemies

These may be

- ☆ predators
- ☆ pathogens
- ☆ parasites

Predators

1. Prey population density effects predator response: Functional response of predator.
2. Prey population effects predator population: Numerical response of predator.

   Predator numbers increase by births and immigration
3. Interaction of prey and predator/predator

   Note that predator numbers will be lower than prey numbers and that there is a time lapse for the predator response.

   With extended time there is an extended cycle of responses of both prey and predator.

The Culprits

- ☆ Predators of insects - ladybird adults and larvae, syrphid larvae, lacewings, praying mantids, dragonflies, Asilids, predatory mites, spiders (account for two thirds of *Heliothis* larvae in cotton), birds
- ☆ Pathogens of insects
  - fungi - can be important in hot, humid summers
  - bacteria - *Bacillus popillae* - milky disease in beetle larvae in pastures
  - *B. thuringieneis* - can be used as biological insecticide
  - Viruses - attack larvae of major orders. Victim's body ruptures releasing viral particles to reinfect others.

- ☆ Parasites of insects
  - **Ectoparasites** live on the outside of the host
  - **Endoparasites** live within the host

Parasitoids are parasitic as larvae, eventually killing their host.

Dispersal and Migration

Insect Migration

Insect migration is the seasonal movement of insects, particularly those by species of dragonflies, beetles, butterflies and moths. The distance can vary with species and in most cases these movements involve large numbers of individuals. In some cases the individuals that migrate in one direction may not return and the next generation may instead migrate in the opposite direction. This is a significant difference from bird migration.

Definition

All insects move to some extent. The range of movement can vary from within a few centimetres for some sucking insects and wingless aphids to thousands of kilometres in the case of other insects such as locusts, butterflies and dragonflies. The definition of migration is therefore particularly difficult in the context of insects.

Migratory behaviour is persistent and straightened-out movement effected by the animal's own locomotory exertions or by its active embarkation on a vehicle. It depends upon some temporary inhibition of station-keeping responses but promotes their eventual disinhibition and recurrence.

This definition disqualifies movements made in the search of resources and which are terminated upon finding of the resource. Migration involves longer distance movement and these movements are not affected by the availability of the resource items. All cases of long distance insect migration concern winged insects.

General Patterns

Migrating butterflies fly within a boundary layer, with a specific upper limit above the ground. The air speeds in this region are typically lower than the flight speed of the insect. These 'boundary-layer' migrants include the larger day-flying insects, and their low-altitude flight is obviously easier to observe than that of most high-altitude windborne migrants.

Many migratory species tend to have polymorphic forms, a migratory one and a resident phase. The migratory phases are marked by their well developed and long wings. Such polymorphism is well known in aphids and grasshoppers. In the migratory locusts, there are distinct long and short-winged forms.

The energetic cost of migration has been studied in the context of life-history strategies. It has been suggested that adaptations for migration would be more valuable for insects that live in habitats where resource availability changes seasonally. Others have suggested that species living in isolated islands of suitable habitats are more likely to evolve migratory strategies. The role of migration in gene flow has also been studied in many species. Parasite loads affect migration. Severely infected individuals are weak and have shortened lifespans. Infection creates an effect known as culling whereby migrating animals are less likely to complete the migration. This results in populations with lower parasite loads.

Orientation

Migration is usually marked by well defined destinations which need **navigation** and **orientation.** A flying insect needs to make corrections for crosswinds. It has been demonstrated that many migrating insects sense wind speed and direction and make suitable corrections. Day-flying insects primarily make use of the sun for orientation; however this requires that they compensate for the movement of the sun. Endogenous time-compensation mechanisms have been proposed and tested by releasing migrating butterflies that have been captured and kept in darkness to shift their internal clocks and observing changes in the directions chosen by them. Some species appear to make corrections while it has not been demonstrated in others.

Most insects are capable of sensing polarized light and they are able to use the polarization of the sky when the sun is occluded by clouds. The orientation mechanisms of nocturnal moths and other insects that migrate have not been well studied, however **magnetic** cues have been suggested in short distance fliers.

Recent studies suggest that migratory butterflies may be sensitive to the Earth's magnetic field on the basis of the presence of magnetite particles. In an experiment on the monarch butterfly, it was shown that a magnet changed the direction of initial flight of migrating monarch butterflies. However this result was not a strong demonstration since the directions of the experimental butterflies and the controls did not differ significantly in the direction of flight.

Lepidoptera

Migration of butterflies and moths is particularly well known. The Bogong moth is a native insect of Australia that is known to migrate to cooler climates. The Madagascan sunset moth (*Chrysiridia rhipheus*) has migrations of up to thousands of individuals, occurring between the eastern and western ranges of their host plant, when they become depleted or unsuitable for consumption.

In southern India, mass migrations of many species occur before monsoons. As many as 250 species of butterflies in India are migratory. These include members of the Pieridae and Nymphalidae. The **Australian painted lady** periodically migrates down the cost of Australia, and occasionally, in periods of strong migration in Australia, migrate to New Zealand.

The **monarch butterfly migrates** from southern Canada to wintering sites in central Mexico where they spend the winter. In the late winter or early spring, the adult monarchs leave the Transvolcanic mountain range in Mexico to travel north. Mating occurs and the females seek out milkweed to lay their eggs, usually first in northern Mexico and southern Texas. The caterpillars hatch and develop into adults that move north, where more offspring can go as far as Central Canada until the next migratory cycle.

Orthoptera

Locusts (*Schistocerca gregaria*) regularly migrate with the seasons.

Short-horned grasshoppers sometime form swarms that will make long flights. These are often irregular and may be related to resource availability and thus not fulfilling some definitions of insect migration. There are however some populations of species such as locusts (*Schistocerca gregaria*) that make regular seasonal movements in parts of Africa; exceptionally, the species migrates very long distances, as in 1988 when swarms flew across the Atlantic Ocean.

Odonata

Dragonflies are among the longest distance insect migrants. Many species of *Libellula, Sympetrum* and *Pantala* are known for their mass migration. *Pantala flavescens* is thought to make the longest ocean crossings among insects, flying between India and Africa on their migrations. Their movements are often assisted by winds.

Coleoptera

Ladybird beetles such as *Hippodamia convergens, Adalia bipunctata* and *Coccinella undecimpunctata* have been noted in large numbers in some places. In some cases, these movements appear to be made in the search for hibernation sites

Local Dispersal and Migration by Insect Pests and their Importance in IPM Strategies

Introduction

The importance of dispersal in insect population dynamics has been largely ignored when formulating integrated pest management (IPM) programs. This is because studies on dispersal by insect pests have focused largely on long-range migration by moths and locusts at high altitudes. For example, the cases of forecasted dispersal reviewed by Pedgley (1993) all involve migration by insects over considerable distances. Migration on a smaller scale has been somewhat ignored. Some feel that long-range dispersal, which we conveniently define as less than 5.0 km, may be rare (Davis 1980; Loxdale *et al.*, 1993), and more emphasis should be placed on short-range dispersal. Short-range migration is certainly as important biologically as long-range dispersal, and is probably more important agronomically.

Laboratory and field experiments concerning migration of agronomic pests on a local scale have largely focused on homopteran insects (Homoptera: Aleyrodidae and Aphididae). Examples are discussed below.

There is no lack of understanding among many academics and growers that migration by pest insects is a phenomenon that impacts crop production. Migration is, however, often omitted as a component of IPM programs at the local level, for two main reasons. To begin, too little is known about the factors that influence migration and dispersal by a particular insect pest. Secondly, testing the impact of this phenomenon on pest populations in agricultural settings is notoriously difficult. This situation is unfortunate, since we require accurate models of insect dispersal to enable realistic forecasting of pest pressure on crops, and this is not possible without detailed knowledge of insect flight behavior and ecology. The interactions between migration and other aspects of pest biology are also critical in developing IPM strategies to deal with mobile pests. For example, in the case of aphids and whiteflies, dispersal of only a few individuals that are insecticide resistant or virus vectors into a crop may give rise to offspring that will cause a significant impact on crop yields.

Definitions

Migration is a widespread phenomenon that has evolved independently in most insect orders (Johnson 1969; Dingle 1972, 1979, 1996; Taylor and Taylor, 1983; Taylor 1986). The definition that best suits the purposes of our research, and that is most consistent with insect behavior we have observed, has been provided by Kennedy (1985).

"Migratory behavior is persistent and straightened-out movement effected by the animal's own locomotory exertions on or by its active embarkation on a vehicle. It depends on some temporary inhibition of station-keeping responses, but promotes their eventual disinhibition and recurrence."

According to Dingle (1996), important characteristics of this definition are:

- ☆ Persistence; that is, the activity of the insect is tenacious so that it carries the migrant beyond its original habitat to a new one where it gathers previously unavailable resources.
- ☆ Dispersal is straightened out; that is, there is a directionality to the dispersal and a minimum of turning and backtracking.
- ☆ Dispersal is undistracted; that is, the insect ignores vegetative cues (cues to resources that could promote growth and maintenance) and keys on other cues, such as sky light, that might be used for activities such as reproduction are employed to support migration.

Flight that has no migratory behavioral component is termed 'trivial flight', and is usually associated with short duration flights between hosts. Crop pest insects that are exhibiting this behavior are attracted to green and yellow wavelengths of light and not to ultraviolet light.In our studies on whitefly flight, insects have been observed in both migratory and trivial flight behavioral phases.

Dispersal is a term that is inclusive of migratory and trivial flight (insects can disperse by either behavior or by random locomotion).Dispersal can be accomplished by populations and not individuals (Southwood 1981). Dispersal may or may not be associated with the characteristics described above.

Dispersal is more simply defined as movement that results in an increase in the mean distance between individuals.

Both migration and dispersal may lead to emigration of a pest from one crop and its eventual immigration into another. Local dispersal by insects may be effected by migratory behaviors, or by short host-seeking flights.

Local Dispersal by Pest Insects

Aphid

Dispersal by insects on a local scale has been extensively studied in aphids. All of these studies use techniques that show that the insects moved from one place to another, but their behavior during dispersal is not known, *i.e.*, they may or may not be engaged in migratory flight. Thus, we cannot categorically state that the insects were all true migrants in the sense of the definition we have provided. Whatever the behavioral states of the insects, they are primarily relocated to new host plants as a result of being carried by the wind. Some of the aphids that take off and have phototactic flight behavior are evidently carried large distances on the wind. These insects may embark on windborne migratory flights that can carry them for great distances, but the ecological importance of this has recently been challenged (Loxdale *et al.*, 1993). They argue that while long distance dispersal is possible, the majority move locally.

Initial Experiments

An early concern, which was pivotal, was whether *B. tabaci* was capable of exhibiting migratory behavior. We had discovered that wingbeat frequency was not significantly correlated with wing loading for these small insects (Byrne *et al.*, 1988). Larger insects seem to employ a strategy similar to other flying animals, *i.e.*, they compensate for high wing loading with higher wingbeat frequencies. The lack of correlation for these two values in whiteflies led to erroneous assumptions concerning the ability of whiteflies to sustain flight. On the other hand, we knew that the internal lipids in adult females of *B. tabaci* represented 40 per cent of their total dry weight(Cohen and Byrne 1992). These compounds are commonly thought to serve as an energy source for insects during flight. For comparison, in the pea aphid, *Acrythosiphum pisum* (Harris), which is known for its migratory ability, lipids account for only 10 per cent of the dry weight. These findings suggested that, despite their small size (33 mg [Byrne *et al.*, 1988]), *B. tabaci* might possess enough energy stores to engage in extended flight.

Laboratory Behavioral Studies

To investigate a series of hypotheses, a vertical flight chamber (Blackmer and Phelan 1991 was used to examine untethered flight of *B. tabaci*. Blackmer and Byrne (1993a) characterized the periodicity, intensity and duration of whitefly flight.

Behaviorally they reaffirmed that there was a dichotomy between groups of whiteflies relative to flight behavior. Some flew for more than 3 h, others flew sporadically, some flew not at all.

Blackmer and Byrne (1993a) also examined how whitefly age and host quality (vegetative vs. senescing) affected flight activity. The propensity to takeoff, proportion exhibiting phototactic orientation, and flight duration varied with whitefly age. Longer-duration flights occurred between three and five days of age. Six days following eclosion little flight took place because, among other things, indirect flight muscles and mitochondria in the thorax had started to deteriorate (Blackmer *et al.*, 1995). Host quality also influenced the timing of flight behavior. Whiteflies reared on senescing plants exhibited greater takeoff rates and initiated longer phototactic flights than individuals reared on vegetative plants.

Laboratory Physiological Studies

It is often assumed that there is a trade-off between migratory capability and reproduction. In migrating individuals, nutritional and energy resources are diverted away from growth and reproduction towards building and maintaining flight muscles and energy resources needed for flight. This allocation of resources into either migration or reproduction has been called the **"oogenesis-flight syndrome"** (Johnson 1969), and has been well documented in alate aphids. We found small, but significant increases, in egg-protein levels were found with longer flights and increasing weights. This was in contrast to the expected (Blackmer *et al.*, 1995).

Lipids are the most suitable materials for storage of energy reserves. Compared to carbohydrates, lipids can supply as much as eight times more energy per unit weight (Beenakkers *et al.*, 1985). In addition, twice as much metabolic water is produced by metabolizing lipids instead of carbohydrates (Downer 1985). A high lipid content associated with migration has been found in many aphids and some bark beetles. In the sweet potato whitefly we found no evidence that lipids are stored for dispersal of less than 3 km. However, flight chamber experiments indicate that in females, lipid content is correlated with flight duration (unpublished data). Thus, lipid reserves could be of more importance for long distance migration.

Understanding the factors leading to production of migratory morphs is important for IPM strategies. Physiological studies can lead to valuable insight into migratory capabilities and will allow us to estimate the effective range of dispersal of insect pests.

Field Studies with Whiteflies

In order to corroborate laboratory observations, field experiments were conducted beginning in 1992 (Byrne *et al.*, 1996). Initial experiments concentrated on the impact of wind flow on the directionality of flight by *B. tabaci.* A 0.73 ha circular field was planted with cantaloupe, *Cucumis melo* L. The area around the field was surveyed for placement of muffin fan traps along a series of 16 transects placed 22.50 apart. These traps are described in Byrne *et al.* (1996). They allowed the capture of living whiteflies.

The evening before each sampling day the melon field was dusted with fluorescent dust. Tracking insects from where they alight to their landing place is possible in the field, and has been done with small insects with some success using the techniques listed separately. The traps were turned on the morning of each sampling day (at approximately 6:00 a.m.). After 3 h, whiteflies marked with dust were collected from the traps. For each sampling date in 1992 the relationship between wind flow and the number of whiteflies captured in traps in the direction opposite the prevailing wind was evaluated. Prevailing wind directions from the northeast (Brown *et al.*, 1995) and trap catch were in agreement for all dates (when the wind was from the northeast > 90 per cent of the whiteflies were caught in traps placed to the southwest of the marked field). This provided empirical evidence that whitefly migration is wind-aided.

Following 1992, traps were located as part of a grid to the southwest of the marked field. Maps of fields and trap locations were digitized using a digitizing tablet. Data were analyzed using the Geostatistical Information System (GIS). The principal feature evident when we examined the data from our field experiments was that whitefly distribution following dispersal was patchy. Almost all of the sampling dates show greatest trap catches of marked insects near the marked field and on the periphery of the grid than in the center of the grid. Pooling data from all sampling dates in 1993 produced a line graph indicating a bimodal distribution. It is evident that the trap catches did not indicate a simple diffusion pattern, a pattern that might be expected if dispersal was passive across featureless topography. This is in contrast to patterns seen by Turchin and Thoeny (1993) who described dispersal by southern pine beetles, *Dendroctonus frontalis* Zimm. using a diffusion model. We believe our pattern may be explained by the existence in whitefly populations of two behaviorally distinct types; trivial flying individuals and migratory individuals as described by Byrne and Houck (1990) and Blackmer and Byrne (1993 a and b). Insects caught in the immediate vicinity of the source field are likely to be trivial fliers, while those caught on the grid perimeter are thought to be migratory individuals. The explanation for the relative paucity of whiteflies caught in the middle of the grid is that migratory whiteflies do not fly towards green vegetation until they have flown for a given period of time (Blackmer and Byrne 1993a).

In our studies, the greatest distance marked whiteflies have been trapped is 2.7 km from a source field (Byrne *et al.*, 1996). Since whiteflies were consistently captured in the most distant traps, the effective migrational range of *B. tabaci* may be considerably greater than this. Future experiments will be conducted with an expanded grid to further define the effective migrational range (*i.e.*, the distance they can travel and still initiate a colony) of *B. tabaci*.

Field research is still ongoing, and these experiments have already revealed a few surprises. For example, we have found that whiteflies were caught in traps placed at the edge of the source field, at over 20 ft above the crop. Traps were in place just before dawn, because the air starts to move at sunrise, and flight is observed as soon as there is sufficient light available to stimulate take-off. Thus, it would seem that the migrants are taking maximum advantage of the low wind speeds to gain height and take them away from the plant boundary layer.This has far-reaching implications for our understanding of whitefly migration and their dispersal on the wind. That whiteflies can take off and ignore plant cues in the field also reinforces our faith that the behaviors observed in the laboratory are those employed by the insects in the field.

Development of Predictive Models

Ideally, our knowledge of insect migration would lead to the development of **simulation models** that serve as an aid to growers in making management decisions. To our knowledge, no such models exist as of yet. This is, in our opinion, partially due to some inherent difficulties in developing and implementing such models.

The "art" of modelling involves reducing a complex biological reality down to a few mathematical statements. For example, one could treat a field of melons, as one unit, with specified size, location, and time of planting. However, the development of whiteflies in a given melon field and their migratory tendencies, are likely to be influenced by the many other factors, like watering regime, amount of fertilizer applied, soil type, or disease incidence. In order to incorporate these factors into the model we need to know the effect of each of these factors on whitefly biology. If one of these factors turned out to be critical for the model, growers would have to spend time (and money) collecting data on this variable. That would severely diminish the value of these models as a management tool.

In addition, one typically has only data available from a small number of successive years for the development of migration models. Thus, the model will be based on a subset of potential environmental conditions. For example, the years for which data were available could be characterized by relatively cool summers. Whether the model thus developed could be extended to years with different temperature regimes can only be examined by collecting data over longer time periods.

Summary

The migratory behaviour of pest insects is an important aspect of their life cycle that has obvious implications for where populations will be exerting pressure on crop production. Our research on whitefly flight behaviour has already yielded information that can be used to assist growers when making decisions concerning crop placement and planting times. The important findings with relevance to IPM are summarized below:

- ☆ Small insect pests can have the capacity for extended flights, controlled by migratory behaviour.
- ☆ Insect dispersal is often in response to both biotic and abiotic conditions that promote migratory behaviour.
- ☆ Migration by insects is an adaptation to a rapidly changing environment. The way in which biotic and abiotic factors influence migration differs widely between species.
- ☆ Wind speed and direction play a significant role in dispersal of small insects.
- ☆ In many insects, migrating individuals are different from non-migrating individuals in physiology and/or morphology.

- ☆ Dispersal by insects acquires additional importance for IPM when they are carrying genes for resistance or are vectors of plant pathogens.
- ☆ They can disperse large distances, but most whiteflies land less than 5.0 km from their original point of take-off.
- ☆ Crop placement and timing should take account of the likely pest pressure from mobile insects.
- ☆ Models can be constructed that help in predicting insect dispersal, but they are only as good as the information they are based on.
- ☆ Knowledge of migration is important for many IPM programs. This does require, however, detailed knowledge of the migratory capacity of each pest insect species.

Seasonality in Insects

Among the studies of life history strategies, a subset focuses on the seasonal timing of life cycle components. When do events like emergence, growth, dispersal and reproduction occur and why at those times? For those who live in temperate zone, seasonal changes are simple facts of life. Cold seasons alternate with warmer seasons, and rainy period may alternate with drier times. Such climatic seasons tends to translate into seasonal activity patterns in living organisms. Birds nest, plants flower and insect abound only at certain times of the year. The higher the altitude, the shorter the growing season- the period in which the physical environment permits the growth and reproduction of plants and animals. The variety of seasonality patterns found, often at one and the same locality, demonstrates the complexity of the matter. Organisms do not respond to the warmer temperature by starting their activity, only to stop when in the fall temperatures drop to lower levels.

Although in the tropics seasonal changes in temperature tend to be minimal or absent, this does not mean that there are no season. Rainy seasons tend to alternate with sometimes one, sometimes two dry seasons each year. Dry seasons may expressed only as "drier season" or they may be entirely absent. Even in such areas, the environment is not necessarily without season.

Classification and Mechanism of Achieving different Seasonality – Diapauses (Quiescence)

Diapause is a suspension of development that can occur at the embryonic, larval, pupal,or adult stage, depending on the species. Diapause is an important means by which insects avoid unfavourable environmental conditions.

Examples of sensitive stage/diapauses periods in various insects

- ☆ Silkworm – Embryonic
- ☆ Grasshopper – Egg
- ☆ Gypsy moth - Late embryonic
- ☆ Southwestern corn borer – late larval
- ☆ Tobacco hornworm – Pupa

- ✰ Cabbage white butterfly - Pupae
- ✰ Colorado potato beetle - Adult
- ✰ Monarch butter fly - Adult

Diapause is of 2 types

1. Facultative Diapause

In some species, diapauses occurs only when indusced by environmental conditions. Diapause affects only some members of a generation resulting in two or more generation in a year (multivoltine).

2. Obligatory diapauses

In some species the diapauses period has become an obligatory part of the life cycle. Every individual

In every generation enter diapauses in obligatory diapauses resulting in only in a single generation (univoltine).

Initiation of Diapause

The period of day light (Photoperiod) is a most important factors in inducing diapauses. Other factors are temperature, food, and age of the parent.

Photoperiod

The day length when 50 per cent of the population has entered diapauses is called the critical day length. There is critical day length around which small differences in photoperiod produced a complete change from non –diapause to diapause development.

Long day insects: Insects entering diapauses when the day length falls below the threshold are called long day insects.

Short day insects: Insects that develop normlley when they are only a few hours of sunlight and that enter diapauses when exposed to loger days are called short day insects. Ex. Bombyx.

Temprature

Temprature also play an important role in the induction of diapauses. In temperate regions, high temperature inhibits and temperature enhances diapausing tendencies. In Bombyx high temperature induces diapauseand low temperature prevents it. There is usually a build up of reserve food substances, particulary in the fat body with a consequent reduction in the proportion of water in the body. In temprate insets, during diapauses there is accumulation of polyhydric alcohols, such as sarbitol and glycerol which can function as cryoprotectants.

Control of diapause

While diapauses in the embryonic stage appear to be regulated (at least in some species) by a diapauses hormone, larval diapauses appear to be controlled by the inhibition of prothoracicotrophic harmone (PTTH) production. This prevents the larvae from moulting and entering pupation. Similarly, the lack of PTTH and ecdysone secretion once pupation has occurred will cause diapauses during this part of development. Diapause is controlled ecdysone in larval and pupal stages its deficiency resulting in arresting of moulting and growth.

Aestivation

Aestivation or **æstivation** (from Latin: *aestas*, summer)is a state of animaldormancy, similar to hibernation, characterized by inactivity and a lowered metabolic rate, that is entered in response to high temperatures and arid conditions.[1] It takes place during times of heat and dryness, the hot dry season, which are often the summer months.

Invertebrate and vertebrate animals are known to enter this state to avoid damage from high temperatures and the risk of desiccation. Both terrestrial and aquatic animals undergo aestivation.

Organisms who aestivate appear to be in a fairly "light" state of dormancy, as their physiological state can be rapidly reversed, and the organism can quickly return to a normal state. A study done on *Otala lactea*, a snail native to parts of Europe and Northern Africa, shows that they can wake from their dormant state within ten minutes of being introduced to a wetter environment. Fossil records show that the act of aestivating may be several hundred million years old.

The primary physiological and biochemical concerns for an aestivating animal are to conserve energy, retain water in the body, ration the use of stored energy, handle the nitrogenous end products, and stabilize bodily organs, cells, and macromolecules. This can be quite a task as hot temperatures and arid conditions may last for months. The depression of metabolic rate during aestivation causes a reduction in macromolecule synthesis and degradation. To stabilize the macromolecules, aestivators will enhance antioxidant defenses and elevate chaperone proteins. This is a widely used strategy across all forms of hypometabolism. These physiological and biochemical concerns appear to be the core elements of hypometabolism throughout the animal kingdom. In other words, animals who aestivate appear to go through nearly the same physiological processes as animals that hibernate.

Molluscs

Gastropoda: Some air-breathing land snails, including species in the genera *Helix, Cernuella, Theba, Helicella, Achatina* and *Otala*, commonly aestivate during periods of heat. Some species move into shaded vegetation or rubble. Others climb up tall plants, including crop species as well as bushes and trees, and will also climb man-made structures such as posts, fences, etc.

The habit of climbing vegetation to aestivate has caused more than one introduced snail species to be declared an agricultural nuisance.

To seal the opening to their shell to prevent water loss, pulmonate land snails secrete a membrane of dried mucus called an epiphragm. In certain species, such as *Helix pomatia*, this barrier is reinforced with calcium carbonate, and thus it superficially resembles an operculum, except that it has a tiny hole to allow some oxygen exchange.

There is decrease in metabolic rate and reduced rate of water loss in aestivating snails like *Rhagada tescorum*, *Sphincterochila boissieri* and others.

Arthropods

Insecta: Lady beetles (*Coccinellidae*) have been reported to aestivate.[4] Mosquitoes also are reported to undergo aestivation.[5]

Crustacea: Many land crabs spend dry seasons in an inactive state at the bottom of their burrows.

Vertebrates

Reptiles and amphibians

Aestivation has been put forward as the most likely explanation why this therapsid cynodont *Thrinaxodon liorhinus* shared its burrow with a temnospondyl amphibian, *Broomistega putterilli*.

Non-mammalian animals that aestivate include North American desert tortoises, crocodiles, and salamanders. Some amphibians (*e.g.* thecane toad and greater siren) aestivate during the hot dry season by moving underground where it is cooler and more humid. The California red-legged frog may aestivate to conserve energy when its food and water supply is low.

The Water-holding Frog has an aestivation cycle. It buries itself in sandy ground in a secreted, water-tight mucus cocoon during periods of hot, dry weather. Australian Aborigines discovered a means to take advantage of this by digging up one of these frogs and squeezing it, causing the frog to empty its bladder. This dilute urine–up to half a glassful–can be drunk. However, this will cause the death of the frog which will be unable to survive until the next rainy season without the water it had stored.

Fish

African lungfish also aestivate.

Mammals

Although relatively uncommon, a small number of mammals aestivate. Animal physiologist Kathrin Dausmann of Philipps University of Marburg, Germany, and coworkers presented evidence in a 2004 edition of *Nature* that the Malagasy fat-tailed dwarf lemur hibernates or aestivates in a small tree hole for seven months of the year. According to the Oakland Zoo in California, East African Hedgehogs are thought to aestivate during the dry season

Hibernation

Insect winter ecology entails the overwinter survival strategies of insects, which are in many respects more similar to those of plants than to many other animals, such as mammals and birds. This is because unlike those animals, which can generate their own heat internally (endothermic), insects must rely on external sources to provide their heat (ectothermic). Thus, insects sticking around in the winter must tolerate freezing or rely on other mechanisms to avoid freezing. Loss of enzymatic function and eventual freezing due to low temperatures daily threatens the livelihood of these organisms during winter. Not surprisingly, insects have evolved a number of strategies to deal with the rigors of winter temperatures in places where they would otherwise not survive.

Two major strategies for winter survival have evolved in the Class Insecta due to their inability to generate significant heat metabolically. The first, migration, is a complete avoidance of the temperatures that pose a threat. If an insect cannot migrate, then it must stay and deal with the cold temperatures in one of two ways. This cold hardiness is separated into two categories, **freeze avoidance** and **freeze tolerance**.

Chapter 9

Biotic Factors: Food is a Limiting Factors for Distribution and Abundance, Nutritional Ecology

Biotic Factors

A) Food

Each insect species has certain nutritional requirements for completion of its life cycle. Under normal conditions there is a happy adjustment between the host and particular species of insect. But in the event of sudden increase in population, the densities of population become too high to be supported by the food available in the area. Hence competition for food as well as space will be there.

According to nutritional requirements, insects are categorized into:

1. **Omnivorous:** Which feed on both plants and animal. *e.g.* Wasps, cockroaches
2. **Carnivorous:** which feed on other animal as parasites and predators. *e.g.* Predators (Lady bird beetles and Mantids)
3. **Herbivorous:** which feed on living plants (crop pests) and these can again be categorized into
   (a) **Polyphagous:** which feed on wide range of cultivated and wild plants. *e.g.* Locusts, grasshoppers
   (b) **Monophagous**: which feed on single species of plants. *e.g.* Rice stem borer
   (c) **Oligophagous**: Which feed on plants of one botanical family. *e.g.* Diamondback moth, Cabbage butterfly.

4. **Saprophagous (Scavengers):** which feed on decaying plants and dead organic matter. *e.g.* Drosophila flies, House flies, scarabaeid beetles.

B) Other Organisms

Include beneficial and harmful insects. Associations of individuals of the same species are known as intra specific relations and it may be beneficial. Such association of two sexes, parental care, associations of social insects etc., phenomenon like overcrowding is harmful since shortage of food and space results. Disease outbreak may occur. Cannibalism may occur. *e.g.* Preying mantids, larvae of Helicoverpa, Tribolium feeds on its own eggs.

Associations of individuals of different species are known as inter-specific relations and these may be beneficial or harmful.

Beneficial Associations

1. **Symbiosis:** Inter relationship between organisms of different species which live in close union without harmful effects are known as symbiosis, each member being known as symbiont.
2. **Commensalism**: One insect is benefited by living on or inside another insect without injuring the other and is known as comensal and it lives on the surplus food or the waste food of its host. *e.g.* Gall forming insects. When the commensal uses its host as a means of transport the phenomenon is termed as phoresy.

e.g., Telonomus beneficiens parasitoid attaches themselves to the anal tufts of female moths of rice stem borer Scirpohaga incertulas (Walker) for their transport. The parasitoid parasitizes freshly laid eggs. iii) Mutualism: When both the symbionts are benefited by the association it is known mutualism *e.g.* Ants and aphids. Termites and flagellates. Harmful associations: Those th ıt live with the expense of other living organisms are parasites and predators.

Parasite: Parasite is one which attaches itself in the body of the other organism either externally or internally for nourishment and shelter at least for a shorter period if not for the entire life cycle. The organism which is attacked by the parasite is called host.

Parasitoid: An insect parasite of arthropod is parasitic only in immature stages, destroys its host in the process of development and free living as an adult or Parasitoid is an insect that feeds on the body of another insect or arthropod during the larval stage of the their life cycle and adult is a free-living insect, no longer dependent on the host. Parasitisation: It is the phenomenon of obtaining nourishment at the expense of the host to which the parasite is attached.

Parasites can be grouped into, depending upon the nature of host, as

1. Zoophagous: That attack animals (Cattle pests)
2. Phytophagous: That attack plants (Crop pests)
3. Entomophagous: That attack insects (Parasitoids and Predators)

Predators and Predatism

A predator is one, which catches and devours smaller or more helpless creatures by killing them in getting a single meal. Insect killed by predator is known as prey.

Insect Predator

It is defined as the one, which is

- Large in size - Active in habits - Capacity for swift movements - Structural adaptations with well developed sense organs to catch the prey - May remain stationary and sedentary - Suddenly seize the pray when it comes with in its reach *e.g.* Antlions - Feed upon large number of small insects every day - May have cryptic colourations and deceptive markings *e.g.* Preying mantids and Robber flies

Nutritional Ecology

Nutritional Ecology of Plant Feeding Insects

I. Insect Nutritional Requirements

A. Carbohydrates: Polyhedroxyl aldehydes and ketones

1 Classifications
   a. Mono-, di-, oligo-, poly- saccharides
   b. Pure, Derived, Conjugated
2. Use: Very similar to mammals
   a. Energy
   b. Vitamin synthesis
   c. Structure: Particularly chitin, Some pigments (protection)
   d. Storage, cold hardiness, pheromone synthesis
   d. Defensive compounds
3. Unutilizable and toxic carbohydrates
   a. Ripe berries often rich in sorbose (unutilizable by insects)
   b. 5-C sugars often toxic: Galactose Arabinose

B. Lipids: Polyunsaturated fatty acids and sterols

1. Use
   a. Structural compounds: cell membranes, cuticle
   b Source of metabolic energy: Linked to flight behavior
   c. Water conservation
2. Absolute requirements
   a. Polyunsaturated fatty acids: C18
   b. Sterols:
      1. Structure
      2. Endocrine system: Ecdysone precursor
      3. Carrier Molecules: Lipoprotein

C. Proteins

1. Uses: extremely broad range
    a. Structure, enzymes, DNA and RNA, storage, etc.
    b. Bioluminescence, Silk
2. Requires dietary Nitrogen
3. Some N containing compounds are toxic: Alkaloids, cyanogenic glycosides, etc.

D. Water

E. Vitamins and Micronutrients: Coenzymes (K,P, Mg, Zn, Fe, Mn, Ca)

II. Insect Nutritional Physiology

A. Digestion and Excretion

1. Carbohydrates
    a Digestion: Simple carbohydrases
    b. Excretion
2. Lipids: Simple hydrolysis
3. Proteins: Enzymes (sensitive to Ph, etc.)

B. Nutritional Indices: Waldbauer 1968

1. Definitions

    Relative Growth Rate (RGR) = Biomass Gain/Original size per day

    Biomass Gain = Ingested - Feces - Metabolism

    Relative Consumption Rate (RCR) = Food Ingested/Original size per day

    Approximate Digestibility (AD) = (Ingested - Feces)/Ingested

    Efficiency of Conversion of Digested Food (ECD) = Biomass G/I-F

    Efficiency of Conversion of Ingested Food (ECI) = Biomass G/I

    RGR = RCR * ECD * AD

    ECI = AD * ECD
2. Homeostasis

C. The dilemma to eating plants: N availability; Implications

III. Variable Distribution of Nutrients

A. Nitrogen

1. Plant Taxa
2. Plant Part

Storage tissues (seeds) > Organs with high turnover rates (Flowers, fruits, leaves, cambium > Saps: (Phloem > xylem sap) > Wood

3. Plant Phenology: Plant Age, Tissue Age
4. Environmental Conditions
58. Interactions with Allelochemicals

B. Water

1. Plant Taxa

   Forbs> Grasses> Shrubs> Trees

2. Plant Part

   Foliage > wood

3. Plant Phenology: Younger tissue has higher water content
4. Environmental Conditions

Variations in Insect Performance on different Plant Parts and Taxa

Approximate digestibilities: tree foliage > forb foliage

ECI's; AD's Seeds, Pollen > Leaves > Wood

IV. Mechanisms for Contending with Low Nitrogen Availability

A. Higher food consumption or development time

B. Specialized digestive systems and symbiotic relationships

   1 Elongated side pouches (caecae) house bacteria, yeasts, protozoa

   2. External symbionts

C. Occasional cannibalism and carnivory

D. Switching among plant parts

E. Regulation of plant chemistry

V. Implications of Variable Nutrient Quality

A. Plant Defense Theory

   1. Induced Defenses
   2. Interactions with Allelochemicals
   3. Plant Stress Theory
   4. Low nutritional quality as a form of defense? Berenbaum

B. Natural and Sustainable Resource Management

   1. Breeding for high quality plants
   2. Fertilization and Irrigation
   3. Acid Precipitation.

Chapter 10
Food Chain: Web and Ecological Succession

Food Chains and Webs

A food chain is the path of food from a given final consumer back to a producer. For instance, a typical food chain in a field ecosystem might be:

grass –> grasshopper –> mouse –> snake –> hawk

Note that the food chain is the path of food from a given final consumer back to a producer typically list a food chain from producer on the left (or at the bottom) to final consumer on the right (or at the top).

The real world, of course, is more complicated than a simple food chain. While many organisms do specialize in their diets (anteaters come to mind as a specialist), other organisms do not. Hawks don't limit their diets to snakes, snakes eat things other than mice, mice eat grass as well as grasshoppers, and so on. A more realistic depiction of who eats whom is called a food web.

It is when we have a picture of a food web in front of us that the definition of food chain makes more sense. We can now see that a food web consists of interlocking food chains, and that the only way to untangle the chains is to trace *back* along a given food chain to its source. The food webs you see here are *grazing food chains* since at their base are producers, which the herbivores then graze on. While grazing food chains are important, in nature they are outnumbered by *detritus-based food chains*. In detritus-based food chains, decomposers are at the base of the food chain, and sustain the carnivores, which feed on them. In terms of the weight (or biomass) of animals in many ecosystems, more of their body mass can be traced back to detritus than to living producers.

Pyramids

The concept of biomass is important. It is a general principle that the further removed a trophic level is from its source (detritus or producer), the less biomass it will contain (biomass here would refer to the combined weight of all the organisms in the trophic level). This reduction in biomass occurs for several reasons:

1. Not everything in the lower levels gets eaten
2. Not everything that is eaten is digested
3. Energy is always being lost as heat

It is important to remember that the decrease in number is best detected in terms or biomass. Numbers of organisms are unreliable in this case because of the great variation in the biomass of *individual* organisms. For instance, squirrels feed on acorns. The oak trees in a forest will always outnumber the squirrels in terms of combined weight, but there may actually be more squirrels than oak trees. Remember that an individual oak tree is huge, weighing thousands of kilograms, while an individual squirrel weighs perhaps 1 kilogram at best. There are few exceptions to the pyramid of biomass scheme. One occurs in aquatic systems where the algae may be both outnumbered and outweighed by the organisms that feed on the algae. The algae can support the greater biomass of the next trophic level only because they can reproduce as fast as they are eaten. In this way, they are never completely consumed. It is interesting to note that this exception to the rule of the pyramid of biomass also is a partial exception to at least 2 of the 3 reasons for the pyramid of biomass given above. While not all the algae are consumed, a greater proportion of them are, and while not completely digestible, algae are far more nutritious overall than the average woody plant is (most organisms cannot digest wood and extract energy from it).

A generalization exists among ecologists that on average, about 10 per cent of the energy available in one trophic level will be passed on to the next; this is primarily due to the 3 reasons given above. Therefore, it is also reasonable to assume that in terms of biomass, each trophic level will weigh only about 10 per cent of the level below it, and 10x as much as the level above it. It also seems, however, that every time I go to measure, test, or model this assumption I run into an inconsistency, so take this generalization with a big grain of salt. Still, it comes in useful In terms of human diet and feeding the world's population, consider this. If we all ate corn, there would be enough food for as many of us as compared to a world where we all eat beef (or chicken, fish, pork, etc.). Another way of looking at it is this. Every time you eat meat, you are taking food out of the mouths of 9 other people, who could be fed with the plant material that was fed to the animal you are eating. Of course, it's not quite that simple, but you get the general idea.

Biological Magnification

Biological magnification is the tendency of pollutants to become concentrated in successive trophic levels. Often, this is to the detriment of the organisms in which these materials concentrate, since the pollutants are often toxic.

Bio-magnification occurs when organisms at the bottom of the food chain concentrate the material above its concentration in the surrounding soil or water. Producers, as we saw earlier, take in inorganic nutrients from their surroundings. Since a lack of these nutrients can limit the growth of the producer, producers will go to great lengths to obtain the nutrients. They will spend considerable energy to pump them into their bodies. They will even take up more than they need immediately and store it, since they can't be "sure" of when the nutrient will be available again (of course, plants don't think about such things, but, as it turns out, those plants, which, for whatever reason, tended to concentrate inorganic nutrients have done better over the years). The problem comes up when a pollutant, such as DDT or mercury, is present in the environment. Chemically, these pollutants resemble essential inorganic nutrients and are brought into the producer's body and stored "by mistake". This is the first step in bio-magnification; the pollutant is at a higher concentration inside the producer than it is in the environment.

The second stage of bio-magnification occurs when the producer is eaten. Remember from our discussion of a pyramid of biomass that relatively little energy is available from one trophic level to the next. This means that a consumer (of any level) has to consume a lot of biomass from the lower trophic level. If that biomass contains the pollutant, the pollutant will be taken up in large quantities by the consumer. Pollutants that bio-magnify have another characteristic. Not only are they taken up by the producers, but they are absorbed and stored in the bodies of the consumers. This often occurs with pollutants soluble in fat such as DDT or PCB's. These materials are digested from the producer and move into the fat of the consumer. If the consumer is caught and eaten, its fat is digested and the pollutant moves to the fat of the new consumer. In this way, the pollutant builds up in the fatty tissues of the consumers. Water-soluble pollutants usually cannot bio-magnify in this way because they would dissolve in the bodily fluids of the consumer. Since every organism loses water to the environment, as the water is lost the pollutant would leave as well. Alas, fat simply does not leave the body.

The "best" example of bio-magnification comes from DDT. This long-lived pesticide (insecticide) has improved human health in many countries by killing insects such as mosquitoes that spread disease. On the other hand, DDT is effective in part because it does not break down in the environment. It is picked up by organisms in the environment and incorporated into fat. Even here, it does no real damage in many organisms (including humans). In others, however, DDT is deadly or may have more insidious, long-term effects. In birds, for instance, DDT interferes with the deposition of calcium in the shells of the bird's eggs. The eggs laid are very soft and easily broken; birds so afflicted are rarely able to raise young and this causes a decline in their numbers. This was so apparent in the early 1960's that it led the scientist Rachel Carson to postulate a "silent spring" without the sound of bird-calls. Her book "*Silent Spring*" led to the banning of DDT, the search for pesticides that would not bio-magnify, and the birth of the "modern" environmental movement in the 1960's. Birds such as the bald eagle have made comebacks in response to the banning of DDT in the US. Ironically, many of the pesticides, which replaced DDT are more dangerous to humans, and, without DDT, disease (primarily in the tropics) claims more human lives.

Summary

In order for a pollutant to bio-magnify, the following conditions must be met:

1. The pollutant must be long-lived.
2. The pollutant must be concentrated by the producers.
3. The pollutant must be fat-soluble.

Human vs. Natural Food Chains

Human civilization is dependent on agriculture. Only with agriculture can a few people feed the rest of the population; the part of the population freed from raising food can then go on to do all the things we associate with civilization. Agriculture means manipulating the environment to favor plant species that we can eat. In essence, humans manipulate competition, allowing favored species (crops) to thrive and thwarting species, which might otherwise crowd them out (weeds). In essence, with agriculture we are creating a very simple ecosystem. At most, it has only three levels - producers (crops), primary consumers (livestock, humans) and secondary consumers (humans). This means that little energy is lost between tropic levels, since there are fewer trophic levels present.

This is good for humans, but what type of "ecosystem" have we created? Agricultural ecosystems have several problems. First, we create *monocultures*, or fields with only one crop. This is simplest for planting, weeding, and harvesting, but it also packs many similar plants into a small area, creating a situation ideal for disease and insect pests. In natural ecosystems, plants of one species are often scattered. Insects, which often specialize on feeding on a particular plant species, have a hard time finding the scattered plants. Without food, the insect populations are kept in check. In a field of corn however, even the most inept insect can find a new host plant with a jump in any direction. Likewise, disease is more easily spread if the plants are in close proximity. It takes lots of chemicals (pesticides) to keep a monoculture going.

Another problem with human agriculture is that we rely on relatively few plants for food. If the corn and rice crops failed worldwide in the same year, we would be hard-pressed to feed everyone (not that we're doing a great job of it now). Natural ecosystems usually have alternate sources of food available if one fails.

A final problem associated with agroecosystems is the problem of inorganic nutrient recycling. In a natural ecosystem, when a plant dies it fall to the ground and rots, and its inorganic nutrients are returned to the soil from which they were taken. In human agriculture, however, we harvest the crop, truck it away, and flush it down the toilet to be run off in the rivers to the ocean. Aside from the water pollution problems this causes, it should be obvious to you that the nutrients are not returned to the fields. They have to be replaced with chemical fertilizers, and that means mining, transportation, electricity, etc. Also, the chemical fertilizers tend to run off the fields (along with soil disrupted by cultivation) and further pollute the water.

Some solutions are at hand, but they bring on new problems, too. No-till farming uses herbicides to kill plants in a field; the crop is then planted through the dead plants without plowing up the soil. This reduced soil and fertilizer erosion, but the herbicides themselves may damage ecosystems. In many areas, sewage sludge is returned to fields to act as a fertilizer. This reduces the need for chemical fertilizers, but still requires a lot of energy to haul the sludge around. Further, if one is not careful, things such as household chemicals and heavy metals may contaminate the sewage sludge and bio-magnify in the crops, which we would then eat.

Interspecific Interactions

The interspecific relationships between various animal species, living together is of a complex nature, particularly when there are large numbers of interacting species in the community. For an understanding of ecological principles, the animals of different species can be divided into two main categories based on the type of interaction they have, *viz.*

(a) Non-predators having horizontal relationships between animals showing common resources such as food and space.

(b) Predators having vertical relationship involving feeding, predation and parasitism.

Predators

A number of mathematical models have been developed to represent the prey – predator interactions. Of these the best known are tose of Lotka (1952),Volterra (1932) and Nicohson and Bailey (1935). Lotka and Volterra showed that prey –predator interactions resulted in oscillatory patterns. According to them, if both prey and predators live in a limited space, the rm of the predator population will be reduced by an amount depending upon its own population and will be increased by an amount depending upon the density of prey. They made assumptions that :

(*i*) The animals moved about at random,

(*ii*) The predators were infallible and their appetites insatiable,

(*iii*) the predator and the prey had a rate of increase conforming to the logistic theory.

Lotka and Volterra gave the following fundamental equations for the prey and the predators living to- gether in a limited space.

$$\frac{\Delta N}{t} = (rm - C'N'')\,N' \text{ - Prey}$$

$$\frac{\Delta N''}{t} = (rm - C''N')\,N'' \text{ - Predator}$$

Where C′ and C″ are constants. C′ is the measure of aptitude of the prey to defend itself. C″ is the measure of effectiveness of the means of offence of a predator.

According to the law of the periodic cycle, the predators increase while the prey is decreasing to a point where this trend becomes reversed. Then for a period, the prey is increasing while the predator decreases until trends are again reversed. This relationship can be expressed by a loop. The height and the loop measure the magnitude of the fluctuations in the population of prey and predators. The point within the loop about which the oscillations occur is called the singular point.

For a predator –prey system, Volterra derived three laws:

(i) The law of *periodic cycle* which predicts that oscillation will occur, depeding upon the increase of predator and prey and the initial relative numbers.

(ii) The law of *conservation of averages* states that provided the intrinsic rate of increase and efficiency of predation are constant, the mean number within each species remain constant, independent of the initial conditions.

(iii) The law of *disturbance of averages*. If two species are uniformly destroyed in proportion of their numbers, the mean number of prey will increase and predators will decrease".

Interspecific interactions

Interaction	Effect on population density
Competition (-/-)	The interaction is detrimental to both species
Predation (+/-) (includes parasitism)	The interaction is beneficial to one species and detrimental to other.
Mutualism (+/+)	The interaction is beneficial to both species
Commensalism (+/0)	One species benefits from the interaction but the other is unaffected

Basic Factors Governing the Interspecific Interaction

Interspecific competition, in ecology, is a form of competition in which individuals of *different* species compete for the same resource in an ecosystem (*e.g.* food or living space). The other form of competition is **intraspecific competition,** which involves organisms of the same species.

If a tree species in a dense forest grows taller than surrounding tree species, it is able to absorb more of the incoming sunlight. However, less sunlight is then available for the trees that are shaded by the taller tree, thus interspecific competition. Leopards and lions can also be in interspecific competition, since both species feed on the same prey, and can be negatively impacted by the presence of the other because they will have less food.

Competition is only one of many interacting biotic and abiotic factors that affect community structure. Moreover, competition is not always a straightforward, direct, interaction. Interspecific competition may occur when individuals of two separate species share a limiting resource in the same area. If the resource cannot support both populations, then lowered fecundity, growth, or survival may result in at least one species. Interspecific competition has the potential to alter populations, communities and the evolution of interacting species. On an individual organism level, competition can occur as interference or exploitative competition.

Direct competition has been observed between individuals, populations and species, but there is little evidence that competition has been the driving force in the evolution of large groups. For example, between amphibians, reptiles and mammals.

Exploitative Competition

Exploitative competition is a form of competition in which one species either reduces or more efficiently uses a resource and therefore depletes the availability of the resource for the other species. Separating the effect of resource use from that of interference is not easy. A good example of exploitative competition is found in **aphid** species competing over the sap in plant phloem. Each aphid species that feeds on host plant sap uses some of the resource, leaving less for competing species. In one study, *Fordinae geoica* was observed to out-compete *F. formicaria* to the extent that the latter species exhibited a reduction in survival by 84 per cent . Another example is driving of bisexual rock lizards of genus*Darevskia* from their natural habitats by a daughter unisexual form; interference competition can be ruled out in this case, because **parthenogenetic** forms of the lizards never demonstrate aggressive behavior.

This type of competition can also be observed in forests where large trees dominate the canopy and thus allow little light to reach smaller competitors living below. These interactions have important implications for the population dynamics and distribution of both species.

Apparent Competition

Apparent competition occurs when two or more species in a habitat affect shared natural enemies in a higher trophic level. If two species share a common predator, for example, apparent competition can exist between the two prey items as they attempt to make themselves less available for consumption.

Scramble Competition

Scramble competition occurs when a resource is inadequate for the needs of all. Each competitor obtains an equal amount, but never the amount it needs, leading to a depression in population growth.

Consequences

Many studies, including those cited previously, have shown major impacts on both individuals and populations from interspecific competition. Documentation of these impacts has been found in species from every major branch of organism. The effects of interspecific competition can also reach communities and can even influence the evolution of species as they adapt to avoid competition. This evolution may result in the exclusion of a species in the habitat, niche separation, and local extinction. The changes of these species over time can also change communities as other species must adapt.

Competitive Exclusion

The competitive exclusion principle states that two species that use the same limiting resource in the same way in the same space and time cannot coexist and must diverge from each other over time in order for the two species to coexist. One species will often exhibit an advantage in resource use. This superior competitor will out-compete the other with more efficient use of the limiting resource. As a result, the inferior competitor will suffer a decline in population over time. It will be excluded from the area and replaced by the superior competitor.

A well documented example of competitive exclusion was observed to occur between Dolly Varden charr (*Salvelinus malma*) and white spotted char (*S. leucomaenis*) in Japan. Both of these species were morphologically similar but the former species was found primarily at higher elevations than the latter. Although there was a zone of overlap, each species excluded the other from its dominant region by becoming better adapted to its habitat over time. In some such cases, each species gets displaced into an exclusive segment of the original habitat. Because each species suffers from competition, natural selection favors the avoidance of competition in such a way.

Niche Differentiation

Niche differentiation is a process by which competitive exclusion leads to differences in resource use. In the previous example, niche differentiation resulted in spatial displacement. In other cases it may result in other changes that also avoid competition. If competition avoidance is achievable, each species will occupy an edge of the niche and will become more specialized to that area thus minimizing competition. This phenomenon often results in the separation of species over time as they become more specialized to their edge of the niche, called niche differentiation. The species do not have to be in separate habitats however to avoid niche overlap. Some species adapt regionally to utilizing different resources than they ordinarily would in order to avoid competition.

There have been several well documented cases in birds where species that are very similar change their habitat use where they overlap. For example, they may consume different food resources or use different nesting habitat or materials. On the Galapagos Islands, finch species have been observed to change dietary specializations in just a few generations in order to utilize limited resources and minimize competition.

In some cases, third party species interfere to the detriment or benefit of the competing species. In a laboratory study, coexistence between two competing bacterial species was mediated by phage parasites. This type of interaction actually helped to maintain diversity in bacterial communities and has far reaching implications in medical research as well as ecology. Similar effects have been documented for many communities as a result of the action of a keystone predator that preys on a competitively superior species.

Local Extinction

Although local extinction of one or more competitors has been less documented than niche separation or competitive exclusion, it does occur. In an experiment involving zooplankton in artificial rock pools, local extinction rates were significantly higher in areas of interspecific competition. In these cases, therefore, the negative effects are not only at the population level but also species richness of communities.

Impacts on Communities

As mentioned previously, interspecific competition has great impact on community composition and structure. Niche separation of species, local extinction and competitive exclusion are only some of the possible effects. In addition to these, interspecific competition can be the source of a cascade of effects that build on each other. An example of such an effect is the introduction of an invasive species to the United States, Purple-loosestrife. This plant when introduced to wetland communities often outcompetes much of the native flora and decreases species richness, food and shelter to many other species at higher trophic levels. In this way, one species can influence the populations of many other species as well as through a myriad of other interactions. Because of the complicated web of interactions that make up every ecosystem and habitat, the results of interspecific competition are complex and site-specific.

Classification of Interspecific Interaction

Biological interactions are the effects organisms in a community have on one another. In the natural world no organism exists in absolute isolation, and thus every organism must interact with the environment and other organisms. An organism's interactions with its environment are fundamental to the survival of that organism and the functioning of the ecosystem as a whole

In ecology, biological interactions can involve individuals of the same species (intraspecific interactions) or individuals of different species (interspecific interactions). These can be further classified by either the mechanism of the interaction or the strength, duration and direction of their effects. Species may interact once in a generation (*e.g.* pollination) or live completely within another (*e.g.*endosymbiosis). Effects range from consumption of another individual (predation, herbivory, or **cannibalism**), to mutual benefit (**mutualism**). Interactions need not be direct; individuals may affect each other indirectly through intermediaries such as shared resources or common enemies.

Interaction Categorized by Effect

Effect on X	*Effect on Y*	*Type of Interaction*
0	0	Neutralism
0	–	Amensalism
+	0	Commensalism
–	–	Competition
+	+	Mutualism
+	–	Antagonism

Some types of relationships listed by the effect they have on each partner. '0' is no effect, '-' is detrimental, and '+' is beneficial.

Terms that explicitly indicate the quality of benefit or harm in terms of fitness experienced by participants in an interaction are listed in the chart. There are six possible combinations, ranging from mutually beneficial through neutral to mutually harmful interactions. The level of benefit or harm is continuous and not discrete, so a particular interaction may have a range from trivially harmful through to deadly, for example. It is important to note that these interactions are not always static. In many cases, two species will interact differently under different conditions. This is particularly true in, but not limited to, cases where species have multiple, drastically different life stages.

Neutralism

Neutralism describes the relationship between two species that interact but do not affect each other. It describes interactions where the health of one species has absolutely no effect whatsoever on that of the other. Examples of true neutralism are virtually impossible to prove. When dealing with the complex networks of interactions presented by ecosystems, one cannot assert positively that there is absolutely no competition between or benefit to either species. However, the term is often used to describe situations where interactions are negligible or insignificant.

Amensalism

Amensalism is an interaction where an organism inflicts harm to another organism without any costs or benefits received by the other. A clear case of amensalism is where sheep or cattle trample grass. Whilst the presence of the grass causes negligible detrimental effects to the animal's hoof, the grass suffers from being crushed. Amensalism is often used to describe strongly asymmetrical competitive interactions, such as has been observed between the Spanish ibex and weevils of the genus *Timarcha* which feed upon the same type of shrub. Whilst the presence of the weevil has almost no influence on food availability, the presence of ibex has an enormous detrimental effect on weevil numbers, as they consume significant quantities of plant matter and incidentally ingest the weevils upon it.

Competition

Competition is a mutually detrimental interaction between individuals, populations or species, but rarely between clades.

Synnecrosis is a particular case in which the interaction is so mutually detrimental that it results in death, as in the case of some parasitic relationships. It is a rare and necessarily short-lived condition as evolution selects against it. The term is seldom used.

Antagonism

In antagonistic interactions, one species benefits at the expense of another. Predation is an interaction between organisms in which one organism captures biomass from another. It is often used as a synonym for carnivory but in its widest definition includes all forms of one organism eating another, regardless of trophic level (*e.g.*, herbivory), closeness of association (*e.g.*, parasitism and parasitoidism) and harm done to prey (*e.g.*, grazing). Intraguild predation occurs when an organism preys upon another of different species but at the same trophic level (*e.g.*, coyotes kill and ingest gray foxes in southern California). Batesian mimicry is also an antagonistic interaction, where one species has evolved to mimic another, to the advantage of the copying species but to the detriment of the species being mimicked.

Ecological Facilitation

The following two interactions can be classed as facilitative. *Facilitation* describes species interactions that benefit at least one of the participants and cause no harm to either. Facilitations can be categorized as mutualisms, in which both species benefit, or commensalisms, in which one species benefits and the other is unaffected. Much of classic ecological theory (*e.g.*, natural selection, niche separation, metapopulation dynamics) has focused on negative interactions such as predation and competition, but positive interactions (facilitation) are receiving increasing focus in ecological research.

Commensalism

Commensalism benefits one organism and the other organism is neither benefited nor harmed. It occurs when one organism takes benefits by interacting with another organism by which the host organism is not affected. A good example is a remora living with a shark. Remoras eat leftover food from the shark. The shark is not affected in the process, as remoras eat only leftover food of the shark, which does not deplete the shark's resources.

Mutualism

Mutualism is an interaction between two or more species, where species derive a mutual benefit, for example an increased carrying capacity. Similar interactions within a species are known as c**o-operation**. Mutualism may be classified in terms of the closeness of association, the closest being symbiosis, which is often confused with mutualism. One or both species involved in the interaction may beobligate, meaning they cannot survive in the short or long term without the other species. Though mutualism has historically received less attention than other interactions such as predation, it is very important subject in ecology. Examples include cleaner fish, pollination andseed dispersal, gut flora, Müllerian mimicry and nitrogen fixation by bacteria in the root nodules of legumes.

Interaction Classified by Mechanism

Symbiosis

The term symbiosis (Greek: living together) can be used to describe various degrees of close relationship between organisms of different species. Sometimes it is used only for cases where both organisms benefit; sometimes it is used more generally to describe all varieties of relatively tight relationships, *i.e.* even parasitism, but not predation. Some even go so far as to use it to describe predation. It can be used to describe relationships where one organism lives on or in another, or it can be used to describe cases where organisms are related by mutual stereotypic behaviors.

In either case, symbiosis is much more common in the living world and much more important than is generally assumed. Almost every organism has many internal parasites. A large percentage of **herbivores** have mutualistic gut fauna that help them digest plant matter, which is more difficult to digest than animal prey. Coral reefs are the result of mutalisms between coral organisms and various types of algae that live inside them. Most land plants and thus, one might say, the very existence of land ecosystems rely on mutualisms between the plants, which fix carbon from the air, and Mycorrhyzal fungi that help in extracting minerals from the ground. The evolution of alleukaryotes (plants, animals, fungi, protists) is believed to have resulted from a symbiosis between various sorts of bacteria: endosymbiotic theory.

Competition

Competition can be defined as an interaction between organisms or species, in which the fitness of one is lowered by the presence of another. Limited supply of at least one resource (such as food, water, and territory) used by both usually facilitates this type of interaction, although the competition may also exist over other 'amenities', such as females for reproduction (in case of male organisms of the same species). Competition is one of many interacting biotic and abiotic factors that affect community structure. Competition among members of the same species is known as **intraspecific competition**, while competition between individuals of different species is known as **interspecific competition**.

Competition between species at the same trophic level of an ecosystem, who have common predators, increases drastically if the frequency of the common predator in the community is decreased by a large margin. The magnitude of competition therefore depends on many factors in the same ecosystem.

According to the competitive exclusion principle, species less suited to compete for resources should either adapt or die out. According to **evolutionary theory**, this competition within and between species, for resources plays a critical role in natural selection.

The Arguments of Cost Benefit Ratios

Abundant empirical evidence shows that biological control, as practiced by professionals is among the most cost effective methods of pest control. Because of its highly positive social and economic benefits, biological control should be among the first pest control tactics to be explored.

Biological control workers must not be indiscriminate in introducing exotic organisms, however. Biological control is a serious endeavor for professionals: it cannot become a panacea for enthusiasts having little of the formal training and understanding of the basis of this discipline. In pest control the rights of society and the environment are increasingly in conflict with private profit. Classical biological control and other forms of natural control, plus other environmentally and economically sound methods must fill the gap. Biological control has the best pest control record and remains a considerable untapped future resource (A. Gutierrez, pers. commun.).

It is difficult to make an analysis of costs and benefits for biological control because the definition "biological control" has been given various meanings (Caltagirone and Huffaker 1980, NAS 1987, Garcia *et al.*, 1988, Gutierrez *et al.*, 1996). Perhaps it is appropriate to distinguish classical and naturally occurring biological control from other methods such as the use of pesticides derived from biological organisms (*e.g.*, *Bacillus thuringiensis* toxins, ryania, pyrethrum, etc.), the use of sterile males, etc.). Gutierrez *et al.* (1991) consider periodic colonization of natural enemies (inundative and inoculative) as an extension of biological control. It is a mistake to call biological control any procedure of pest control that involves the use or manipulation of a biological organism or its products as was done by Reichelderfer (1979, 1981, 1985). Reichelderfer's contribution has been to show how economic theory applies to an analysis of the economic benefits of augmentative releases of biological control agents, and in this sense the arguments are similar to those for estimating the benefits of using pesticides or any other control method.

In this discussion of economic gains, the discipline of biological control as an applied activity, concerns itself with the introduction and conservation of natural enemies that become, or are essential components of self-generating systems in which the interacting populations (principally predator/prey or parasitoid/host) are regulated. In biological control of pests the manipulated organisms include predators, parasitoids, pathogens and competitors. No judgments are made concerning the value of other procedures, except to note those which encourage environmentally safe and economically sound approaches. Biological control of pests has been implemented worldwide, in environments that are climatically, economically and technologically diverse (Clausen 1978). The net benefits derived from this tactic as a whole are difficult to quantify with any degree of accuracy. However, the considerable number of cases that were successful, and continue to be so, and the fact that no environmental damage has been detected in the great majority of them make this tactic a very desirable one. Nevertheless, the classical biological control approach (introduction of exotic natural enemies) has been challenged on the basis of possible negative effect on native organisms. For example, Howarth (1983) proposed that in Hawaii the introduction of some natural enemies has adversely affected the native fauna, and that to restore the ecological situation by removal of these organisms is nearly impossible. This points to the vexing aspect of possible environmental risk in using exotic biological control agents (Legner *1985, 1986*). It has been accepted that these organisms, when introduced according to restrictions established by regulatory agencies (Animal and Plant Health Inspection Service in the United States) are considered to pose no environmental hazard. Routinely, risk is recognized when considering candidate natural enemies to control weeds. A comprehensive discussion on this aspect of biological control is given by Turner (1985), and Legner (1986a,b).

The biological impact of exotic biological control agents on target pests is difficult to assess and few cases have been thoroughly documented (Luck *et al.*, 1988), making economic analysis difficult. Even more demanding would be to include in the equation the monetary value of the side effects as referred to by Howarth (1983) and the positive ones (*e.g.*, the benefit that society derives from the reduction in or the elimination of the use of objectionable pesticides) as a result of the introduction of an effective natural enemy.

Biological Control From Naturally Occurring Organisms

The economic benefits of naturally occurring biological control have been repeatedly demonstrated in those cases where secondary pests became unmanageable as a result of overuse of chemical pesticides to control primary pests. The rice brown plant hopper, *Nilaparvata lugens*, in southeastern Asia continued to be a pest as a result of it overcoming the new varieties' resistance and the use of pesticides to control it.

Host plant resistance may be overcome by natural selection of new biotypes of phytophages in the field in less than seven years. The rice brown planthopper is a product of the green revolution wherein the increased prophylactic use of pesticide destroyed its natural enemies and caused the secondary outbreak of this pest. Recognition of this problem recently led to the banning of many pesticides in rice in Indonesia. This prohibition has resulted in no losses in rice yields. This often causes these pests to become more important than the original target pests. These examples substantiate the benefits of naturally occurring natural enemies in controlling pests. Furthermore, these benefits are largely free of cost, unless special procedures are required to either augment or reintroduce them.

Estimation of Benefits and Costs of Classical Biological Control

The costs of a classical biological control project (C) may be calculated easily. One simply sums the cost of the base line research, the cost of foreign exploration, shipping, quarantine processing, mass rearing, field releases and post release evaluation. The last cost must be evaluated judiciously as pursuing academic interests may push these costs beyond those required by the practical problem at hand.

DeBach (1974) gave a rough estimate of the cost of importing natural enemies at the University of California. He commented that he had imported several natural enemies into various countries with resulting impressive practical successes where the cost had been less than $100 per species. In other cases the cost may run much higher, but he believed not more than a few thousand dollars per entomophagous species at most. These tentative costs suggest that some classical biological control projects may be very inexpensive, but others may cost more because of the biological and other complexities encountered. Also, the efficiency of the organization involved may cause costs to vary considerably, and the cost of the biological control efforts on a per organization, per country, or worldwide basis must include the cost of fruitless efforts. Like any other tactic, biological control must record not only its successes but also failures (Ehler and Andrés 1983). A monetary loss due to a failure may still provide a scientific gain in knowledge which is usually unmeasurable. Such knowledge may be applied positively in future efforts with a consequent savings of cost.

Once establishment and dispersal in the new environment is obtained in classical biological control, no further costs for this natural enemy are incurred unless new biotypes are introduced. Other uses of natural enemies may involve repeated releases of natural enemies in the field or glasshouse. These costs are analogous to the cost of pesticide applications.

The environmental costs of biological control derived from the possible suppression or eradication of native species by introduced exotic natural enemies (Howarth 1983, Turner 1985) could be included in a benefit/cost analysis if some monetary value could be placed on them. More often than not such factors cannot be accurately priced in much the same way that increased cancer risks due to the use of some pesticides cannot be priced.

Biological Control Benefit Computation is a more difficult task. One of the most successful, and historically the first, case of biological control in California was the control of the cottony cushion scale, *Icerya purchasi*, by the imported natural enemies *Rodolia cardinalis* and *Cryptochaetum iceryae*. In 1889-1889, when these natural enemies were imported to California at the cost of a few hundred dollars, the young citrus industry was at the verge of collapse because of the scale. One year later shipments of oranges from Los Angeles County had increased three-fold . What figures should we use to determine the benefits of such a program? Obviously the benefits continue to accrue to the present. In 1889 there was no other effective way to control the scale even though it is possible that some of the modern chemical pesticides could control it today. So is the yearly benefit the full net value of the citrus crop (assuming the uncontrolled pest could destroy all of the crop and many of the trees as well), or the total cost of using an effective pesticide? Should we include the benefits of introducing these natural enemies from California to 26 other countries, in 23 of which the scale was completely controlled? Whichever method is chosen, the benefits of this project are vast but undocumented.

Much more difficult are those cases were partial noneconomic control occurs: the natural enemy becomes established, regulates the population of the target species to a lower level, but not low enough as to have economic significance. It is conceivable that in cases like these the natural enemies may make it easier to implement a more effective, complementary control tactic (*e.g.*, IPM). The effects of biological interactions are complex and they are often influenced by other factors including weather, and the beneficial effects of the natural enemy may not be obvious. When the results of biological control are clear-cut, increased production and increased land values may be only part of the equation, as enhanced environmental and health effects may also occur but may go undocumented. The basis for a comparison between the situation prior and after establishment of biological control must further consider the changing real value of money over time, changing markets for the commodity involved, and the dynamics of land use. Enhanced yield may be due to reduced pest injury, but also to reduction in diseases the pest may vector.

Benefits which are easiest to estimate are those to the agricultural sector. Because of the permanent nature of biological control, the net benefits (II) [*i.e.*, benefit (B) - costs (C)] corrected for the present value of money using the discount rate (1 + @)-1 accrue over t years (i = 1,...,t). Note that @ is the interest rate of price of money.

t

$II = Z\ (B_{i-C}i)/(1 + @)i$

$1=1$

[Z = summation sign]

Gross revenue (B) to the grower equals P (Y-DN(1-E)) with P being price, Y the maximum possible yield, D the damage rate per pest N, and E the efficacy of the biological control. In reality, D is a function of N (*i.e.*, D(N(1-E))), but for simplicity we assume that D is a constant. In fact, the benefit of biological control for the ith year is $B_i = PDN_iE$, and in the extreme may equal PY.

There are also recent cases of successful biological control where the benefits are just as impressive but an economic analysis has not been conducted. The control of three Palearctic cereal aphids over the wheat growing regions of South America reduced the pesticide load on the environment causing direct enhancement of yields. New wheat varieties were being developed at the time, but their yield potential had not been stabilized. Thus it is not possible to assess the maximum contribution of the biological control effort. But if as a result of the establishment of natural enemies there was a saving of one application of pesticide per annum the total savings in Argentina, Brazil and Uruguay on 8,996,000 ha. of wheat alone (FAO 1987) would be substantial, especially if it is contrasted with the cost of the biological control component, which has been estimated at less than $300,000.

Justification of Need for Biological Control

The question is then why do we feel the need to make economic justifications for biological control? Why hasn't biological control been more widely supported worldwide? Economists would call this a market failure, because the users of pesticides do not pay for long term consequences of pesticide use and hence may ignore environmentally safer alternatives (Regev 1984). But there are also problems of perception, for as Day (1981) assessed in his review of the acceptance of biological control as an alternative for control of alfalfa weevil in the northeastern United States: "At first, the general opinion was that biological insect control was outmoded, because it had not been effective in the east in decades, and it was not likely to be competitive with synthetic insecticides or the newer synthetic chemicals such as pheromones, chemosterilants, attractants and hormones." Thus, biological control was not appreciated as competitive with newer technologies and it was not considered modern. The recent over selling of bioengineering solutions for crop protection can also be added to the list of reasons why classical biological control is not currently strongly supported.

The technologically advanced countries the advocates of biological control, compared to those promoting predominantly the use of chemical pesticides, are much fewer in number, generally have sparser resources and have a more difficult educational task. It requires great educational skills, financial resources and personal dedication to effectively convey the necessary information in order to enable growers to make educated decisions about pest control. The processes of biological control are not visible to the majority of agriculturists, and with rare exception its benefits become part of the complicated biology that is absorbed in the business of crop production, and is quickly forgotten by old and new clients alike. On rare occasions the biological and economic success was so dramatic, as occurred with Klamath weed in California, that the generations four decades later is aware of the history of the control. The problem is also increasing in developing countries as modern agrotechnology displaces traditional methods, and they too become dependent on pesticides for the control of pests. To combat this problem the United Nations sponsored project on rice in southeastern Asia headed by P. E. Kenmore has set as its goal the training of millions of rice farmers on how to recognize the organisms responsible for the natural control of rice pests. Thus, perceptions of the seriousness of a pest control problem often determine whether an environmentally sound alternative is selected.

Biological Control and Pesticide Use

In a free market economy individual growers make their own pest control decisions, and purveyors of alternatives such as pesticides have the right to market them in accordance with state laws. Under such a system, the perceptions of the problem by growers and the marketing skills of those proposing alternative solutions often dictate how well biological control is adopted in the field.

In evaluating the effectiveness of chemical control or augmentative release of natural enemies, economists traditionally look at the balance of revenues (B(x)) = the value of the increase in yield attributable to using x units of the control measure (*e.g.*, pesticide or augmentation) minus the out-of-pocket cost (C(x)) of causing x units of the control measure. Only infrequently are the social costs (S(x)) associated with the control measure included. For augmentative releases of natural enemies and biological control, S(x) is usually zero. The benefit function is usually assumed to be concave from below and the cost per unit of x constant. The net benefit (II) function should be:

$$II = B(x) - C(x)$$

The optimal solution to this function occurs when dB/dx = dC/dx, hence the optimal quantity of x to use is x_1 when S(x) is excluded, but is x_2 when included? If the cost per unit of x used increases with x, costs rise rapidly and less pesticide (x_3) is optimal. Unfortunately, the social or external costs of pesticides in terms of pollution, health and environmental effects are seldom included in the grower's calculations because there is no economic incentive to do so. In contrast, augmentative releases of natural enemies also engender ongoing costs, but they are environmentally safe and may be more economical than pesticide use. Prime examples of the successful use of this method are the highly satisfactory control of pests in sugarcane in Latin America.

Conservation of natural enemies for control of pests such as *Lygus* bugs on cotton in the San Joaquin Valley in California and in other crops elsewhere (DeBach 1974) often yields superior economic benefits than does insecticidal control (Falcon *et al.*, 1971). In such cases the ill advised use of chemical pesticides (x) may induce damage resulting in additional pest control costs and, at times, lower yields (Gutierrez *et al.*, 1979). With naturally occurring biological control and economically viable classical biological control (BC), the costs of other pest control tactics and social costs often become zero, and the whole of society obtains the maximum benefits: the natural and biological controls supplant other methods of control and may solve the problem permanently.

It is therefore important how a grower understands risk which determines how much he will be willing to pay for pest control to minimize that risk. Adding the social cost of pesticide use to the cost of pesticides narrows the gap between unrealistically perceived risk and the real risk to profits. Taxing pesticide users to fund biological control efforts would be a socially responsible way to fund permanent solutions for pest problems

Analysis of Successes

Parasitoids have been argued to be better than predators as biological control agents. Because a predaceous larva consumes many host individuals during its lifetime and a parasitoid but one host, it might appear that a predator is inherently more destructive and thus makes a better biological control agent. However, analysis of the 139 species of entomophagous insects imported and established in the United States as of 1967 showed that 113 were parasitoids and 26 predators. This ratio has remained similar into the 1990's. Roughly twice as many successes in biological control have resulted from parasitoid introduction in the United States. However, about four times as many on the world scene.

The apparent superiority of parasitoids is the subject of contemporary debate and research. This may only reflect the fact that parasitoids have received the greatest amount of attention in terms of the number of species introduced and the number subjected to field analyses.

There are of course many other countries reporting successful cases of biological control. Many of these are represented by only one or two successes that resulted largely from trans-shipments of biological control agents of proven value following their initial successful employment in other countries. Four insect pests that have been controlled in this manner in various countries are:

A. Cottony-cushion scale controlled by the ***Rodolia*** (Vedalia) beetle in 55 countries following its initial success in California.

B. Woolly apple aphid controlled by ***Aphelinus mali*** in 42 of 51 countries into which it was introduced following its initial success in New Zealand.

C. White peach scale controlled by ***Prospaltella berlesei*** in 5 countries following its initial success in Italy.

D. Citrus blackfly controlled by ***Eretomocerus serius*** in 9 countries following its initial success in Cuba.

Pest Groups

Further analysis reveals that 55 per cent of the 107 pest species brought under some measure of biological control through 1960 belong to the Homoptera, nearly 40 per cent of which are scale insects. 20 per cent of the pests are Lepidoptera; 17 per cent are Coleoptera, while 8 per cent belong to other taxa.

Natural Enemy Groups

Because a majority of successes have involved coccids, it follows that a large proportion of the natural enemies involved in biological control success have been natural enemies of scale insects:

Hymenoptera– Encyrtidae and Aphelinidae

Coleoptera–Coccinellidae

Chapter 11

Interspecific Competition: Lotka-Volterra-Model

Introduction

Interspecific competition refers to the competition between two or more species for some limiting resource. This limiting resource can be food or nutrients, space, mates, nesting sites– anything for which demand is greater than supply. When one species is a better competitor, interspecific competition negatively influences the other species by reducing population sizes and/or growth rates, which in turn affects the population dynamics of the competitor. The Lotka-Volterra model of interspecific competition is a simple mathematical model that can be used to understand how different factors affect the outcomes of competitive interactions.

Importance

Competitive interactions between organisms can have a great deal of influence on species evolution, the structuring of communities (which species coexist, which don't, relative abundances, etc.), and the distributions of species (where they occur). Modeling these interactions provides a useful framework for predicting outcomes.

The impacts of interspecific competition on populations have been formalized in a mathematical model called the **Competitive Lotka–Volterra equations**, which creates a theoretical prediction of interactions. It combines the effects of each species on the other. These effects are calculated separately for the first and second population respectively:

$$\frac{dN_1}{dt} = r_1 N_1 \frac{K_1 - N_1 - \alpha_{12} N_2}{K_1}$$

$$\frac{dN_2}{dt} = r_2 N_2 \frac{K_2 - N_2 - \alpha_{21} N_1}{K_2}$$

In these formulae, N is the population size, t is time, K is the carrying capacity, r is the intrinsic rate of increase and α is the competition coefficient. The results show the effect that the other species has on the species being calculated. The results can be graphed to show a trend and possible prediction for the future of the species. One problem with this model is that certain assumptions must be made for the calculation to work. These include the lack of **migration** and constancy of the carrying capacities and competition coefficients of both species. The complex nature of ecology determines that these assumptions are rarely true in the field but the model provides a basis for improved understanding of these important concepts.

Question

Under what circumstances can two species coexist? Under what circumstances does one species out compete another?

Variables

- N population size
- t time
- K carrying capacity
- r intrinsic rate of
- α competition coefficient

Methods

The logistic equation below models a rate of population increase that is limited by intraspecific competition (*i.e.*, members of the same species competing with one another).

$$\frac{dN}{dt} = rN\frac{(K-N)}{K}$$

The first term on the right side of the equation (*rN*, the intrinsic rate of increase [*r*] times the population size [*N*]) describes a population's growth in the absence of competition. The second term ([*K-N*]/*K*) incorporates intraspecific competition, or density-dependence, into the model, and takes a value between 0 and 1. As population size (*N*) approaches carrying capacity (*K*), the numerator (*K-N*) becomes smaller but the denominator (*K*) stays the same and the second term decreases. The addition of this term describes a rate of population growth that slows down as population size increases, until the population reaches its carrying capacity. In other words, the growth curve described by the logistic equation is sigmoidal, and the rate of growth depends on the density of the population.

The logistic equation can be modified to include the effects of interspecific competition as well as intraspecific competition. The Lotka-Volterra model of interspecific competition is comprised of the following equations for population 1 and population 2, respectively:

$$\frac{dN_1}{dt} = r_1 N_1 \frac{(K_1 - N_1 - \alpha_{12} N_2)}{K_1}$$

$$\frac{dN_2}{dt} = r_2 N_2 \frac{(K_2 - N_2 - \alpha_{21} N_1)}{K_2}$$

The big difference (other than the subscripts denoting populations 1 and 2) is the addition of a term involving the competition coefficient, α. The competition coefficient represents the effect that one species has on the other: α_{12} represents the effect of species 2 on species 1, and α_{21} represents the effect of species 1 on species 2 (the first number of the subscript always refers to the species being affected). In the first equation of the Lotka-Volterra model of interspecific competition, the effect that species 2 has on species 1 (α_{12}) is multiplied by the population size of species 2 (N_2). When α_{12} is < 1 the effect of species 2 on species 1 is less than the effect of species 1 on its own members. Conversely, when α_{12} is > 1 the effect of species 2 on species 1 is greater than the effect of species 1 on its own members. The product of the competition coefficient, α_{12}, and the population size of species 2, N_2, therefore represents the effect of an equivalent number of individuals of species 1, and is included in the intraspecific competition, or density-dependence, term. The $\alpha_{21}N_1$ term in the second equation is interpreted in the same way.

To understand the predictions of the model it is helpful to look at graphs that show how the size of each population increases or decreases when we start with different combinations of species abundances (*i.e.*, low *N*1 low*N*2, high *N*1 low *N*2, etc.). These graphs are called state-space graphs, in which the abundance of species 1 is plotted on the *x*-axis and the abundance of species 2 is plotted on the *y*-axis. Each point in a state-space graph represents a combination of abundances of the two species. For each species there is a straight line on the graph called a zero isocline.

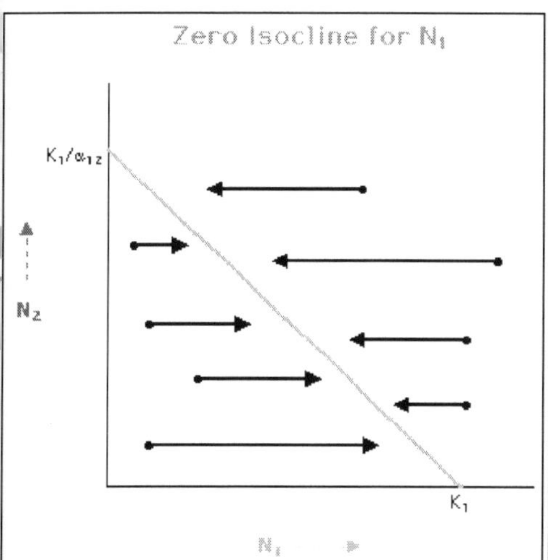

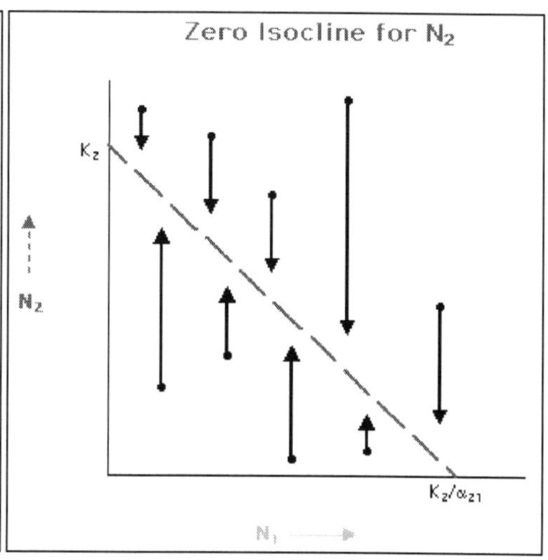

Any given point along, for example, species 1's zero isocline represents a combination of abundances of the two species where the species 1 population does not increase or decrease (the zero isocline for a species is calculated by setting *dN/dt*, the growth rate, equal to zero and solving for *N*). The two graphs below show the zero isoclines for species 1 (left, solid yellow line) and species 2 (right, dashed pink line).

Note that the zero isoclines divide each graph into two parts. Below and to the left of the isocline the population size increases because the combined abundances of both species are less than the carrying capacity of the one, while above and to the right the population size decreases because the combined abundances are greater than the carrying capacity. For the graph of the isocline of species 1, the isocline intersects the graph on the *x*-axis when *N*1 reaches its carrying capacity (K_1) and no individuals of species 2 are present. The isocline intersects the graph on the *y*-axis at K_1/α_{12}, when the carrying capacity of species 1 is filled by the equivalent number of individuals of species 2 and no individuals of species 1 are present. The intersections of the isocline for species 2 are essentially the same, but on different axes.

These two graphs illustrate what happens to a population when it is below or above its isocline, but they only account for one isocline at a time. The following four graphs include both species' isoclines, and illustrate the possible outcomes of interspecific competition depending on where each species' isocline lies in relation to the other. In each graph, the solid yellow line represents the isocline of species 1, and the dashed pink line represents the isocline of species 2. The thick black arrows represent the joint trajectory of the two populations, and the thinner colored arrows indicate the trajectories of the individual populations.

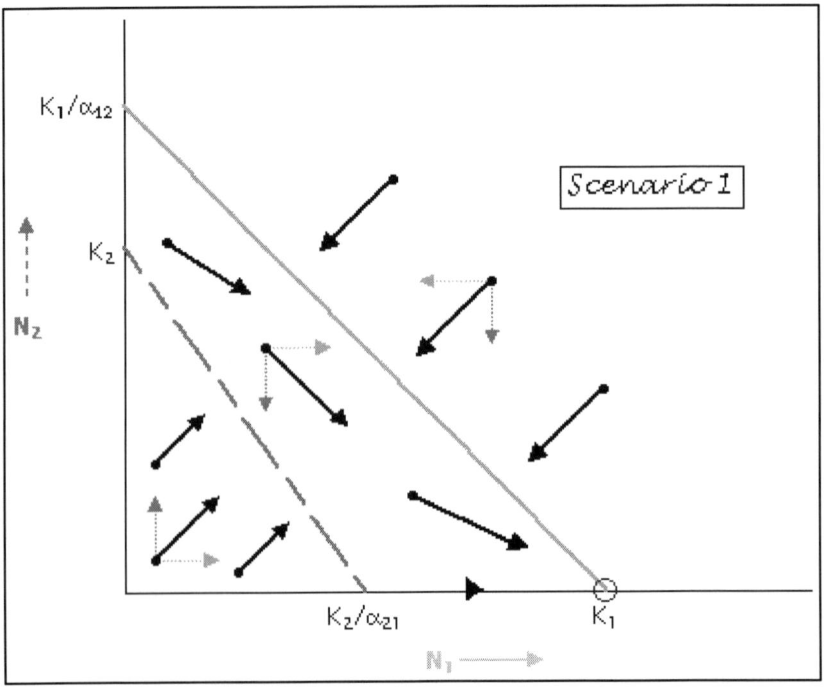

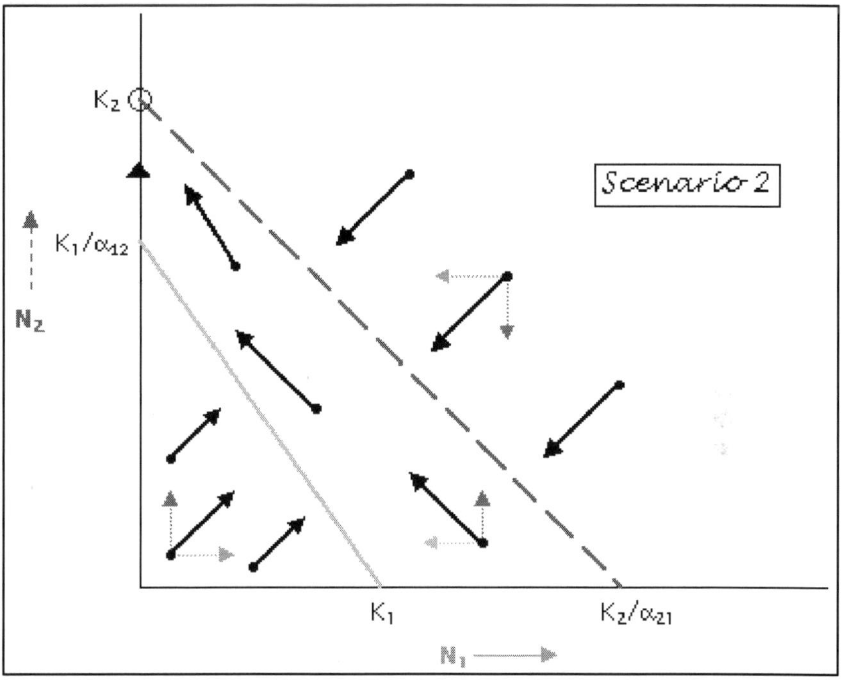

Interpretation

The first scenario is one in which the isocline for species 1 is above and to the right of the isocline for species two. For any point in the lower left corner of the graph (*i.e.*, any combination of species abundances), both populations are below their respective isoclines and both increase. For any point in the upper right corner of the graph, both species are above their respective isoclines and both decrease. For any point in between the two isoclines, species 1 is still below its isocline and increases, while species 2 is above its isocline and decreases. The joint movement of the two populations (thick black arrows) is down and to the right, so species 2 is driven to extinction and species 1 increases until it reaches carrying capacity (K_1). The open circle at this point represents a stable equilibrium. In this scenario, species 1 always outcompetes species 2, and is referred to as the competitive exclusion of species 2 by species 1.

The second scenario is the opposite of the first; the isocline of species 2 is above and to the right of the isocline for species 1. This graph can be interpreted in much the same way as the previous one, except that the joint trajectory of the two populations when starting in between the isoclines is up and to the left. In this case species 2 always outcompetes species 1, and species 1 is competitively excluded by species 2.

In the third scenario, the isoclines of the two species cross one another. Here, the carrying capacity of species 1 (K_1) is higher than the carrying capacity of species 2 divided by the competition coefficient (K_2/α_{21}), and the carrying capacity of species 2 (K_2) is higher than the carrying capacity of species 1 divided by the competition

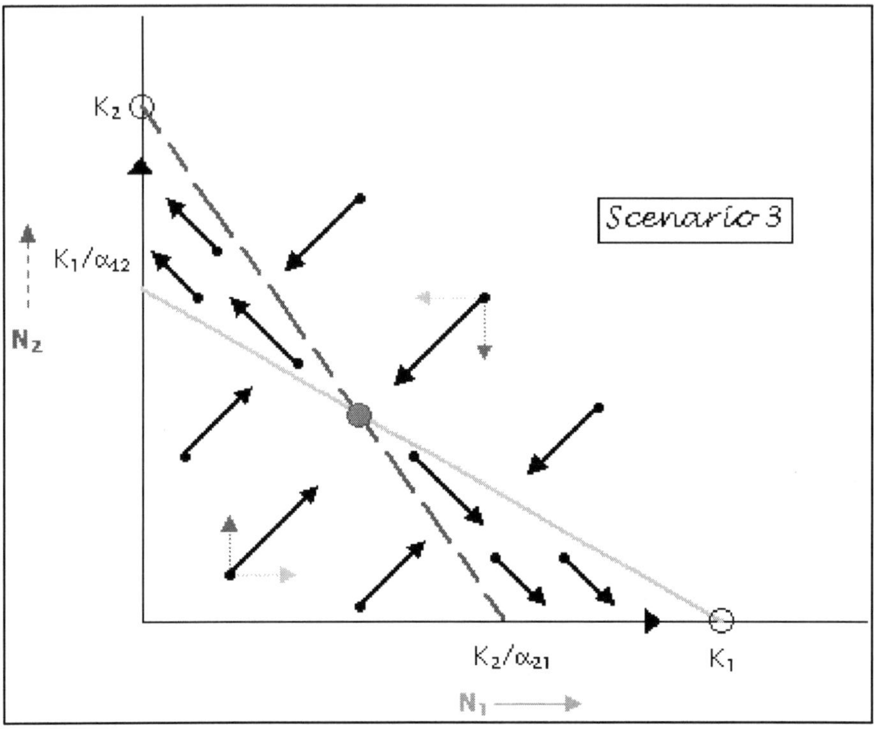

Scenario 3
K2
K1/α12
N2
K2/α21
K1
N1

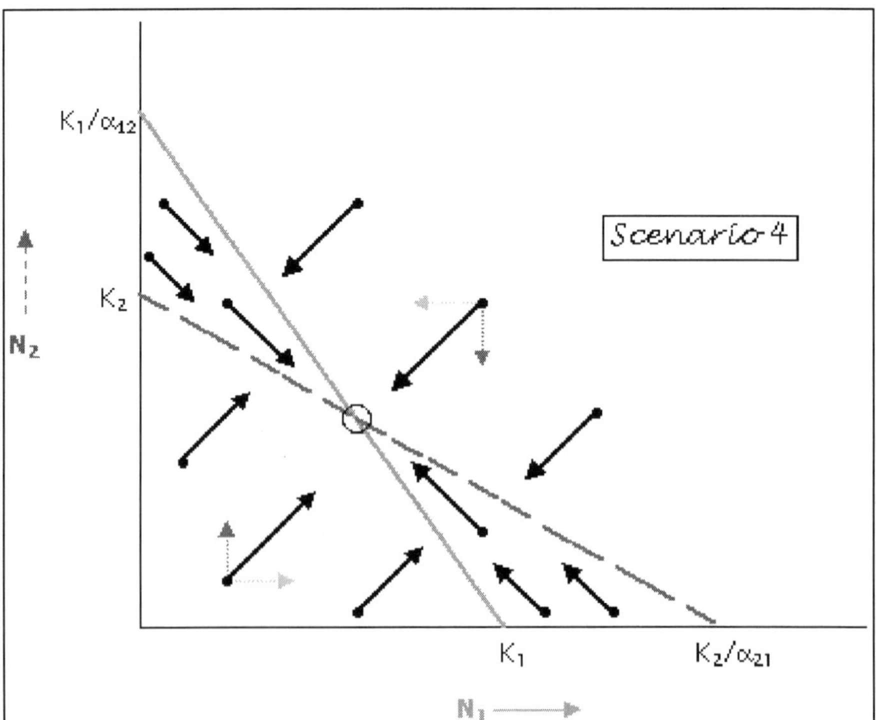

Scenario 4
K1/α12
K2
N2
K1
K2/α21
N1

coefficient (K_1/α_{12}). Below both isoclines and above both isoclines the populations increase or decrease as in the first two scenarios, and there is an unstable equilibrium point (closed circle) where the isoclines intersect. For points above the dashed pink line (species 2 isocline) and below the solid yellow line (species 1 isocline), the outcome is the same as in the first scenario: competitive exclusion of species 2 by species 1. On the other hand, for points above the solid yellow line (species 1 isocline) and below the dashed pink line (species 2 isocline), the outcome is the same as in the second scenario: competitive exclusion of species 1 by species 2. The two stable equilibrium points are again represented by open circles. In this scenario, the outcome depends on the initial abundances of the two species.

Finally, in the fourth scenario we can see that the isoclines cross one another, but in this case both species' carrying capacities are lower than the other's carrying capacity divided by the competition coefficient. Again, below both isoclines the populations increase and above both isoclines the populations decrease. In this case, however, when the populations of the two species are between the isoclines their joint trajectories always head toward the intersection of the isoclines. Rather than outcompeting one another, the two species are able to coexist at this stable equilibrium point (open circle). This is the outcome regardless of the initial abundances.

Conclusions

The Lotka-Volterra model of interspecific competition has been a useful starting point for biologists thinking about the outcomes of competitive interactions between species. The assumptions of the model (*e.g.*, there can be no migration and the carrying capacities and competition coefficients for both species are constants) may not be very realistic, but are necessary simplifications. A variety of factors not included in the model can affect the outcome of competitive interactions by affecting the dynamics of one or both populations. Environmental change, disease, and chance are just a few of these factors.

Chapter 12

Concept of Niche: Ecological Homologous Competitive Exclusion

Prey-Predators Interactions – Defence Mechanism against Predators/Parasitoids – Evolution of Mimicry, Colouration, Defense Mechanism against Predators/Parasitoids

Concept of Niche

Ecological niche is a particular combination of physical factors (microhabitat) and biotic relations (role) required by a species for its life activities and continued existence in a community. It was Joseph Grinnell of the University of California who for the first time used the term *ecological niche*. He considered ecological niche the ultimate distributive or spatial unit occupied by just one species, or subspecies, to which that species is held by structural and instinctive limitations such as climatic factors, kind and amount of food, suitable nesting-sites, and cover. At about the same time Charles Elton in England defined the *niche* in more functional terms as an animal's place in the biotic environment, its relation to food and enemies. The present-day concept of niche is an elaboration of these basic ideas, with an emphasis on all the relation of the organism to both the physical and biotic factors of its environment. Hutchinson (1957) defined *fundamental niche* and *realized niche*. The environmental variable range within which a species can survive and reproduce is a hyper-volume or n-dimensional abstract volume is the *fundamental niche*. The *realized niche* of a species is restricted to that portion of its fundamental niche to which it is better adapted. The variety of species living together in different niches gives that ecosystem *species diversity*.

Characteristics of the Niche

Microhabitat

The physical features of the microhabitat – substratum, space, and microclimate – are basic determinants of whether a particular niche can be occupied by a given species.

Diurnation

When two or more species are competing for the same resources of a single habitat, this competition is reduced or eliminated if one species makes use of these resources at different times of the day, or in a different season than the other. Time then becomes a dimension in which species occurring in the same community may differ in niche.

Aspection

Species of certain grasshoppers (*Arphia sulphurea* and *A. xanthoptera*) occupy similar niches except *A. sulphurea* over winters in the nymph stage, reaching maturity from April to July. The other species over winters in the egg stage and requires longer time to mature in the spring and occur from late July through November. Predatory arthropods segregate out seasonally in temperate forests and fields: the large species of spiders predominate from March to May, the medium-sized ones predominate from May and June and the carabid beetles from July through September.

Shelter and Vegetation

Animals require shelter or cover as a protection against unfavourable weather and enemies. Vegetation is an important source of cover/shelter for animals. Shelter is required either to avoid too much solar insolation or to escape the notice of predators. Depending upon the purpose for the cover or shelter is used we term it as escape cover, nesting cover, winter cover and roosting cover. If the type of cover required by a species is missing, that species will not occur even if all other conditions are favourable.

Food and Predators

Since most organisms select their food from that most easily available to them, it is usually more important in characterizing their niches to indicate the type or characteristics of food consumed and the stratum or microhabitat from which it is obtained. Two or more species may feed together in the same community but be segregated from each other because they search for their food from different plant species, from different parts of the same plant or they take different food from the same parts of same plants. For describing the position of a species occupies in the food chains of a community, it is necessary to indicate not only the kinds of food eaten but also what species prey upon it and the manner in which they do so. Some parasitoids are quite specific as to the kind of host in which to deposit eggs. If a parasitoid, specific to a host species, is present in a niche, he simple presence of that parasitoid will determine the success with which the host species will compete for and fully occupy the niche.

Behaviour Adjustments

Somehow an animal must get into a favourable niche out of the multitude of niches available to it. Some animals accomplish this in a passive manner through the random dispersal of spores, eggs, or larval stages, some of which reach the favourable niche/locations to mature and survive. Where two species with similar niche requirements come into competition, one species must possess better adaptation to it if it gains full possession of the niche to the exclusion of the other species. Moreover, pre-adaptation or possession of suitable adaptations also appears a necessary pre-requisite for a species to invade a new niche or habitat.

Inheritance

The fact that all individuals of a species behave almost in an identical manner in many of their activities indicates that these behavioral patterns are in some form passed on or inherited from one generation to the next. The transmission of behavioral patterns is often by genetic mechanisms, but can also be by tradition (training of young by parents, young imitating parents, conditioning or imprinting). Some parasitoid wasps lay their eggs only in the same kind of host-larvae, as they themselves became conditioned to during their early growth and feeding.

Ecological Homologous

Convergent evolution describes the independent evolution of similar features in species of different lineages. Convergent evolution creates analogous structures that have similar form or function, but that were not present in the last common ancestor of those groups. The cladistic term for the same phenomenon is homoplasy, from Greek for *same form*. The recurrent evolution of flight is a classic example of convergent evolution. Flying insects, birds, and bats have all evolved the capacity of flight independently. They have "converged" on this useful trait.

Functionally similar features arising through convergent evolution are termed *analogous*, in contrast to homologous structures or traits, which have a common origin, but not necessarily similar function. The British anatomist Richard Owen was the first scientist to recognise the fundamental difference between analogies and homologies. Bat and pterosaur wings constitute an example of analogous structures, while the bat wing is homologous to human and other mammal forearms, sharing an ancestral state despite serving different functions. The opposite of convergent evolution is divergent evolution, whereby related species evolve different traits. On a molecular level, this can happen due to random mutation unrelated to adaptive changes.

Convergent evolution is similar to, but distinguishable from, the phenomena of parallel evolution. Parallel evolution occurs when two independent but similar species evolve in the same direction and thus independently acquire similar characteristics– for instancegliding frogs have evolved in parallel from multiple types of tree frog.

Causes

In morphology, analogous traits will often arise where different species live in similar ways and/or similar environment, and so face the same environmental factors.

When occupying similar ecological niches (that is, a distinctive way of life) similar problems lead to similar solutions.

In biochemistry, physical and chemical constraints on mechanisms cause some active site arrangements to independently evolve multiple times in separate enzyme superfamilies.

Significance

Convergence has been associated with Darwinian evolution in the popular imagination since at least the 1940s. For example, Elbert A. Rogers argued: "If we lean toward the theories of Darwin might we not assume that man was [just as] apt to have developed in one continent as another?"

In his book *Wonderful Life*, Stephen Jay Gould argues that if the tape of life were re-wound and played back, life would have taken a very different course. Simon Conway Morris disputes this conclusion, arguing that convergence is a dominant force in evolution, and given that the same environmental and physical constraints are at work, life will inevitably evolve toward an "optimum" body plan, and at some point, evolution is bound to stumble upon intelligence, a trait presently identified with at least primates, corvids, andcetaceans.

Convergence is difficult to quantify, so progress on this issue may require exploitation of engineering specifications (as of wing aerodynamics) and comparably rigorous measures of "very different course" in terms of phylogenetic (molecular) distances.

Distinction

Convergent evolution is a topic touched by many different fields of biology, many of which use slightly different nomenclature. This section attempts to clarify some of those terms.

Cladistic Definition

In cladistics, a homoplasy or a homoplastic character state is a trait (genetic, morphological etc.) that is shared by two or more taxa because of convergence, parallelism or reversal. Homoplastic character states require extra steps to explain their distribution on a most parsimonious cladogram. Homoplasy is only recognizable when other characters imply an alternative hypothesis of grouping, because in the absence of such evidence, shared features are always interpreted as similarity due to common descent. Homoplasious traits or changes (derived trait values acquired in unrelated organisms in parallel) can be compared with synapomorphy (a derived trait present in all members of a monophyletic clade), autapomorphy (derived trait present in only one member of a clade), or apomorphies, derived traits acquired in all members of a monophyletic clade following divergence where the most recent common ancestor had the *ancestral* trait (the ancestral trait manifesting in paraphyletic species as a plesiomorphy).

Re-evolution vs. Convergent Evolution

In some cases, it is difficult to tell whether a trait has been lost then re-evolved convergently, or whether a gene has simply been 'switched off' and then re-enabled later. Such a re-emerged trait is called an atavism. From a mathematical standpoint, an unused gene (selectively neutral) has a steadily decreasing probability of retaining potential functionality over time. The time scale of this process varies greatly in different phylogenies; in mammals and birds, there is a reasonable probability of remaining in the genome in a potentially functional state for around 6 million years.

Parallel vs. Convergent Evolution

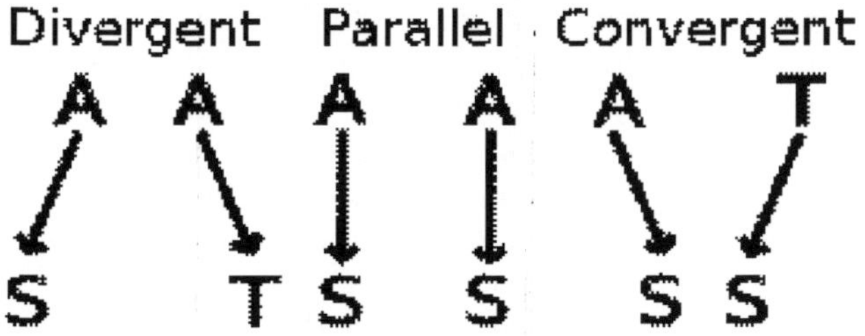

Evolution at an amino acid position. In each case, the left-hand species changes from incorporating alanine (A) at a specific position within a protein in a hypothetical common ancestor deduced from comparison of sequences of several species, and now incorporates serine (S) in its present-day form. The right-hand species may undergo divergent, parallel, or convergent evolution at this amino acid position relative to that of the first species.

For a particular trait, proceeding in each of two lineages from a specified ancestor to a later descendant, parallel and convergent evolutionary trends can be strictly defined and clearly distinguished from one another. However the cutoff point for what is considered convergent and what is considered parallel evolution is assigned somewhat arbitrarily. When two species are similar in a particular character, evolution is defined as parallel if the ancestors were also similar and convergent if they were not. However, this definition is somewhat murky. All organisms share a common ancestor more or less recently, so the question of how far back to look in evolutionary time and how similar the ancestors need to be for one to consider parallel evolution to have taken place is not entirely resolved within evolutionary biology. Some scientists have argued parallel evolution and convergent evolution are more or less indistinguishable from one another. Others have argued that we should not shy away from the gray area and that there are still important distinctions between parallel and convergent evolution.

When the ancestral forms are unspecified or unknown, or the range of traits considered is not clearly specified, the distinction between parallel and convergent evolution becomes more subjective. For instance, the striking example of similar placental and marsupial forms is described by Richard Dawkins in The Blind Watchmaker as a case of convergent evolution, because mammals on each continent had a long evolutionary history prior to the extinction of the dinosaurs under which to accumulate relevant differences. Stephen Jay Gould describes many of the same examples as parallel evolution starting from the common ancestor of all marsupials and placentals. Many evolved similarities can be described in concept as parallel evolution from a remote ancestor, with the exception of those where quite different structures are co-opted to a similar function. For example, consider *Mixotricha paradoxa*, a microbe that has assembled a system of rows of apparent cilia and basal bodies closely resembling that of ciliates but that are actually smaller symbiont micro-organisms, or the differently oriented tails of fish and whales. On the converse, any case in which lineages do not evolve together at the same time in the same ecospace might be described as convergent evolution at some point in time.

The definition of a trait is crucial in deciding whether a change is seen as divergent, or as parallel or convergent. In the image above, note that, since serine and threonine possess similar structures with an alcohol side-chain, the example marked *"divergent"* would be termed *"parallel"* if the amino acids were grouped by similarity instead of being considered individually. As another example, if genes in two species independently become restricted to the same region of the animals through regulation by a certain transcription factor, this may be described as a case of parallel evolution – but examination of the actual DNA sequence will probably show only divergent changes in individual base-pair positions, since a new transcription factor binding site can be added in a wide range of places within the gene with similar effect.

A similar situation occurs considering the homology of morphological structures. For example, many insects possess two pairs of flying wings. In beetles, the first pair of wings is hardened into wing covers with little role in flight, while in flies the second pair of wings is condensed into small halteres used for balance. If the two pairs of wings are considered as interchangeable, homologous structures, this may be described as a parallel reduction in the number of wings, but otherwise the two changes are each divergent changes in one pair of wings.

Similar to convergent evolution, evolutionary relay describes how independent species acquire similar characteristics through their evolution in similar ecosystems, but not at the same time (dorsal fins of sharks and ichthyosaurs).

Example

Insect Mouthparts

Insect mouthparts show many examples of organs which can be used to study convergent evolution in the context of form and function. The mouthparts of different insect groups consist of a set of homologous organs, specialised to the function of dietary intake of that insect group (which can be experimentally quantified). Convergent Evolution of many groups of insects led from original biting-chewing mouthparts to different derived function types. They build a proboscis at flower-visiting insects for example, which are able to ingest food very efficiently or biting-sucking mouthparts, showing different function mechanisms at different groups of blood-sucking insects.

Prey-predators Interactions

Types of Insect Predators

Birds, Fishes, Amphibians, Mammals, other insects and arthropods.

Types of Insect Defenses

- ☆ **Primary Defense-** Camouflage, Mimesis, Mimicry
- ☆ **Secondary Lines of Defense-** Mechanical Defenses, Chemical Defenses Behavioral, Individual and Group
- ☆ **Tertiary Defenses**

  Primary Defense
- ☆ Defense that reduces the probability of encounter by a natural enemy

  Crypsis-Visual deception to avoid predator encounters

  Camouflage- Camouflage-resembling the general background

  Mimesis (masquerade)

  Masquerade- resembling an object that is a particular feature of the environment, and is uninteresting to the predator

Mimicry

Components of a Mimicry System

- ☆ **Aposematic**- Model- Gives the basic warning signal
- ☆ **Mimic-**An organism that parasitizes the signaling system through deluding the observer.
- ☆ **Observer-**Observer-Selecting agent (predator) which associates the warning signal given by the model with some sort of pain that it wishes to avoid
- ☆ **Mimicry-** Chemically defended insects tend to be apparent, or obvious, to their predators
- ☆ **Aposematism-** Aposematism-warning of unpalatability, particularly using color, but sound or light can also be used.

Predators often learn to avoid toxically defended predators by their aposematic coloration or patterns patterns.Undefended insects can mimic these colors or patterns as a defense mechanism without expending the energy to actually make chemicals.

Insects have a wide variety of predators including birds, reptiles, amphibians, mammals, carnivorous plants, and other arthropods. The great majority (80–99.99 per cent) of individuals born do not survive to reproductive age, with perhaps 50 per cent of this mortality rate attributed to predation. In order to deal with this ongoing escapist battle, insects have evolved a wide range of defense mechanisms. The only restraint on these adaptations is that their cost, in terms of time and energy, does not exceed the benefit that they provide to the organism in question. The further that a feature tips the balance towards beneficial, the more likely that selection will act upon the trait, passing it down to further generations. The opposite also holds true;

defenses that are too costly will have a slim to zero chance of being passed down. Examples of defenses that have withstood the test of time include hiding, escape by flight or running, and firmly holding ground to fight as well as producing chemicals and social structures that help prevent predation.

One of the best known modern examples of the role that evolution has played in insect defenses is the link between melanism and the peppered moth (*Biston betularia*). Peppered moth evolution over the past two centuries in England has taken place, with darker morphs becoming more prevalent over lighter morphs so as to reduce the risk of predation. However, its underlying mechanism is still debated

Hiding

The camouflaged stick insect ***Medauroidea extradentata*** Walking sticks (order Phasmatodea), many katydid species (family Tettigoniidae), and moths (order Lepidoptera) are just a few of the plethora of insects that have evolved specialized **cryptic** morphology. This adaptation allows them to effectively hide within their environment because of a resemblance to the general background or an inedible object. When an insect looks like an inedible or inconsequential object in the environment that is of no interest to a predator, such as leaves and twigs, it is said to display **mimesis**, a form of **crypsis**.

Insects may also take on different types of **camouflage**, another type of cypsis. These include resembling a uniformly colored background as well as being light below and dark above, or **coundershaded**. Additionally, camouflage is effective when it results in patterns or unique morphologies that disrupt outlines so as to better merge the individual into the background.

Cost and Benefit Perspective

Butterflies (order Lepidoptera) are a good example of the balancing act between the costs and benefits associated with defense. In order to take off, butterflies must have a thorax temperature of 36–40 °C (97–104 °F). This energy is derived both internally through muscles and externally through picking up solar radiation through the body or wings. When looked at in this light, cryptic coloration to escape from predators, markings to attract conspecifics or warn predators (**aposematism**), and the absence of color to absorb adequate solar radiation, all play key roles in survival. Only when these three affairs are in balance does the butterfly maximize its fitness.

Mimicry

Mimicry is a unique form of defense and describes when a species resembles another recognized by natural enemies, giving it protection against predators. The superficial resemblance among mimics does not necessarily denote common ancestry. Mimicry works if and only if predators are able to learn from eating distasteful species. It is a three part system that involves a model species, a mimic of that species, and a predatory observer that acts as a selective agent. If learning is to be successful, then all models, mimics, and predators must co-exist, a notion feasible within the context of geographic sympatry.

Mimicry is divided into two parts, **Batesian mimicry** and **Müllerian mimicry**.

Batesian Mimicry

In Batesian mimicry, an **aposematic** inedible model has an edible mimic. Automimics are individuals that, due to environmental conditions, lack the distasteful or harmful chemicals of conspecifics, but are still indirectly protected through their visibly identical relatives. An example can be found in the Plain Tiger (*Danaus chrysippus*), a non-edible butterfly, which is mimicked by multiple species, the most similar being the female Danaid Eggfly (*Hypolimnas misippus*).

Müllerian Mimicry

In Müllerian mimicry, a group of species benefit from each other's existence because they all are warningly colored in the same manner and are distasteful. The best examples of this phenomenon can be found within the Heliconius butterfly genus. Like in Batesian mimicry, the mimics are not necessarily related, although they obviously are in Heliconius.

Behavioral Responses

Behavioral responses to escape predation include burrowing into substrate and being active only through a restricted period of the day. Furthermore, insects may feign death, a response termed **thanatosis**. Beetles, particularly **weevils**, are especially fond of this sort of acting. Bright colors may also be flashed underneath cryptic ones. A startle display occurs when prey takes advantage of these markings after being discovered by a predator.The striking color pattern, which often includes eyespots, is intended to evoke prompt enemy retreat. Better formed eyespots seem to result in better deterrence.

Mechanical Defenses

Insects have had millions of years to evolve a variety of mechanical defenses. Perhaps the most obvious is the cuticle. Although its main role lies in support and muscle attachment, when extensively hardened by the cross-linking of proteins and chitin, or sclerotized, the cuticle acts as a first line of defense. Additional physical defenses include modified mandibles, horns, and spines on the tibia and femur. When these spines take on a main predatory role, they are termed **raptorial.**

Some insects uniquely create retreats that appear uninteresting or inedible to predators. This is the case in caddisfly larvae (order *Trichoptera*) which encase their abdomen with a mixture of natural materials like leaves, twigs, and stones.

Autotomy

Autotomy, or the shedding of appendages, is also used to distract predators, giving the prey a chance to escape. This highly costly mechanism is regularly practiced within stick insects (order Phasmatodea) where the cost is accentuated by the possibility that legs can be lost 20 per cent of the time during molting. Harvestmen (order Opiliones) also use autotomy as a first line of defense against predators.

Chemical Defenses

Unlike pheromones, **allomones** harm the receiver at the benefit of the producer. This grouping encompasses the chemical arsenal that numerous insects employ. Insects with chemical weaponry usually make their presence known through aposematism. **Aposematism** is utilized by non-palatable species as a warning to predators that they represent a toxic danger. Additionally, these insects tend to be relatively large, long-lived, active, and frequently aggregate. Indeed, longer-lived insects are more likely to be chemically defended than short lived ones, as longevity increases apparency.

Throughout the arthropod, and insect, realm, however, chemical defenses are quite unevenly distributed. There is great variation in the presence and absence of chemical arms among orders and families to even within families. Moreover, there is diversity among insects as to whether the defensive compounds are obtained intrinsically or extrinsically.Many compounds are derived from the main food source of insect larvae, and occasionally adults, feed, whereas other insects are able to synthesize their own toxins.

In reflex bleeding, insects dispel their blood, hemolymph, or a mixture of exocrine secretions and blood as a defensive maneuver. As previously mentioned, the discharged blood may contain toxins produced within the insect source or externally from plants that the insect consumed. Reflexive bleeding occurs in specific parts of the body; for example, the beetle families Coccinellidae (ladybugs) and Meloidae bleed from the knee joints.

Classification of Chemical Defenses

Gullan and Cranston have divided chemical defenses into two classes. Class I chemicals irritate, injure, poison, or drug individual predators. They can be further separated into immediate or delayed substances, depending on the amount of time it takes to feel their effects. Immediate substances are encountered topographically when a predator handles the insect while delayed chemicals, which are generally contained within the insect's tissues, induce vomiting and blistering. Class I chemicals include bufadienolides, cantharidin, cyanides, cardenolides, and alkaloids, all of which have greater effects on vertebrates than on other arthropods. In fact, the most frequently encountered defensive compounds in insects are alkaloids.

In contrast to Class I chemicals, Class II chemicals are essentially harmless. They stimulate scent and taste receptors so as to discourage feeding. They tend to have low molecular weight and are volatile and reactive, including acids, aldehydes, aromatic ketones, quinones, and terpenes. Furthermore, they may be aposematic, indicating through odors the presence of chemical defenses. The two different classes are not mutually exclusive, and insects may use combinations of the two.

Pasteels, Grégoire, and Rowell-Rahier also grouped chemical defenses, albeit in a different manner. Weaponry is partitioned into chemical compounds that are truly poisonous, those that restrict movement, and those that repel predators. True poisons, essentially Class I compounds, interfere with specific physiological processes or act at certain sites. Repellents are similar to those classified under Class II as they irritate the chemical sensitivity of predators. Impairment of movement and sense organs is achieved through sticky, slimy, or entangling secretions that act mechanically rather than chemically. This last grouping of chemicals has both Class I and Class II properties. As with Class I and Class II compounds, these three categories are not mutually exclusive, as some chemicals can have multiple effects.

Examples: An Assassin Bug

Some insects inject their venom directly into their victims, as is the case with the over 5,000 species in the family Reduviidae, the assassin bugs and thread-legged bugs. Interestingly, the assassin bug species Platymeris rhadamanthus is capable of spitting venom up to 30 cm. The saliva of this insect, loaded with a mixture of at least six proteins including excess amounts of protease, hyaluronidase, and phospholipase causes intense local pain, vasodilation, and edema.

Cockroaches

Many cockroach species (order Blattodea) have mucus-like adhesive secretions on their posterior. Although not as effective against vertebrates, these secretions foul the mouths of invertebrate predators, increasing the chances of the cockroach escaping.

Termites

The majority of termite soldiers secrete a rubberlike and sticky chemical concoction that serves to entangle enemies, called a Fontanellar gun, and it is usually coupled with specialized mandibles. In nasute species of termites (contained within the subfamily Nasutitermitinae), the mandibles have receded. This makes way for an elongated, syringic nasus capable of squirting liquid glue. When this substance is released from the frontal gland reservoir and dries, it becomes sticky and is capable of immobilizing attackers. It is highly effective against other arthropods, including spiders, ants, and centipedes.

Among termite species in the Apicotermitinae that are soldierless or where soldiers are rare, mouth secretions are commonly replaced by abdominal **dehiscence**. These termites contract their abdominal muscles, resulting in the fracturing of the abdominal wall and the expulsion of gut contents. Because abdominal dehiscence is quite effective at killing ants, the noxious chemical substance released is likely contained within the termite itself.

Ants

Venom is the defense of choice for many ants (family Formicidae). It is injected from an **ovipositor** that has been evolutionarily modified into a stinging apparatus. These ants release a complex venom mixture that can include histamine. Within the Formicinae subfamily, the stinger has been lost and instead the poison gland forcibly ejects the fluid of choice, formic acid. Some carpenter ants (genus *Camponotus*) also have mandibular glands that extend throughout their bodies. When these are mechanically irritated, the ant commits suicide by exploding, spilling out a sticky, entangling substance.

The subfamily Dolichoderinae, which also does not possess a stinger, has a different type of defense. The anal gland secretions of this group rapidly polymerize in air and serve to immobilize predators.

Leaf Beetles

The larvae of leaf beetles often employ fecal shields, masses of feces that they carry on their bodies to repel predators. More than just a physical barrier, the fecal shield contains excreted plant volatiles that can serve as potent predator deterrents.

Wasps

Ant attacks represent a large predatory pressure for many species of wasps, including the *Polistes versicolor*. These wasps possess a gland located in the VI abdominal sternite (van de Vecht's gland) that is primarily responsible for making an ant repellent substance. Tufts of hair near the edge of the VI abdominal sternite store and apply the ant repellent, secreting the ant repellent through a rubbing behavior.

Collective Defense in Social Insects

Many chemically defended insect species take advantage of clustering over solitary confinement. Among some insect larvae in the orders **Coleoptera** and **Hymenoptera,** cycloalexy is adopted. Either the heads or ends of the abdomen, depending on where noxious compounds are secreted, make up the circumference of a circle. The remaining larvae lie inside this defensive ring where the defenders repel predators through threatening attitudes, **regurgitation**, and biting.

Termites (order Isoptera), like eusocial ants, wasps, and bees, rely on a caste system to protect their nests. The evolution of fortress defense is closely linked to the specialization of soldier mandibles. Soldiers can have biting-crushing, biting-cutting, cutting, symmetrical snapping, and asymmetrical snapping mandibles. These mandibles may be paired with frontal gland secretion, although snapping soldiers rarely utilize chemical defenses. Termites take advantage of their modified mandibles in **phragmosis**, which is the blocking of the nest with any part of the body; in this case of termites, nest entrances are blocked by the heads of soldiers.

Alarm **pheromones** warn members of a species of approaching danger. Because of their altruistic nature, they follow the rules of **kin selection.** They can illicit both aggregational and dispersive responses in social insects depending on the alarm caller's location relative to the nest. Closer to the nest, it causes social insects to aggregate and may subsequently produce an attack against the threat. The *Polistes canadensis*, a primitively eusocial wasp species, will emit a chemical alarm substance at the approach of a predator, which will lower their nestmates' thresholds for attack, and even attract more nestmates to the alarm. The colony is thus able to rise quickly with its sting chambers open to defend its nest against predators. In nonsocial insects, these compounds typically stimulate dispersal regardless of location.

Chemical alarm systems are best developed in aphids and treehoppers (family Membracidae) among the nonsocial groups. Alarm pheromones take on a variety of compositions, ranging from terpenoids in aphids and termites to acetates, an alcohol, and a ketone in honeybees to formic acid and terpenoids in ants.

Immunity

Insects, like nearly every other organism on Earth, are subject to infectious diseases caused by viruses, bacteria, fungi, protozoa, and **nematodes**. These encounters can either kill the insect or greatly weaken it. Insects protect themselves against these detrimental microorganisms in two ways. Firstly, the body-enveloping chitin cuticle, in conjunction with the tracheal system and the gut lining, serve as major physical barriers to entry. Secondly, hemolymph itself plays a key role in repairing external wounds as well as destroying foreign organisms within the body cavity. Interestingly, insects, along with having passive immunity, also show evidence of acquired immunity.

Social insects additionally have a repertoire of behavioural and chemical 'border-defences' and in the case of the ant, groom venom or metapleural gland secretions over their cuticle.

Phenotypic Plasticity

Phenotypic plasticity is the capacity of a single genotype to exhibit a range of phenotypes in response to variation in the environment. For example, in Nemoria arizonaria caterpillars, the cryptic pattern changes according to season and is triggered by dietary cues. In the spring, the first brood of caterpillars resembles oak catkins, or flowers. By the summer when the catkins have fallen, the caterpillars discretely mimic oak twigs. No intermediate forms are present in this species, although other members of the *Nemoria* genus, such as *N. darwiniata*, do exhibit transitional forms.

In social insects such as ants and termites, members of different castes develop different phenotypes. For example, workers are normally smaller with less pronounced mandibles than soldiers. This type of plasticity is more so determined by cues, which tend to be non-harmful stimuli, than by the environment.

Phenotypic plasticity is important because it allows an individual to adapt to a changing environment and can ultimately alter their evolutionary path. It not only plays an indirect role in defense as individuals prepare themselves physically to take on the task of avoiding predation through camouflage or developing collective mechanical traits to protect a social hive, but also a direct one. For example, cues elicited from a predator, which may be visual, acoustic, chemical, or vibrational, may cause rapid responses that alter the prey's phenotype in real time.

Chemical Defenses

Many insects are equipped to wage chemical warfare against their enemies. In some cases, they manufacture their own toxic or distasteful compounds. In other cases, the chemicals are acquired from host plants and sequestered in the hemolymph or body tissues. When threatened or disturbed, the noxious compounds may be released onto the surface of the body as a glandular ooze, into the air as a repellent volatile, or aimed as a spray directly at the offending target.

Defensive chemicals typically work in one of four ways:

1. **Repellency** – a foul smell or a bad taste is often enough to discourage a potential predator. Stink bugs, for example, have specialized exocrine glands located in the thorax or abdomen that produce foul-smelling hydrocarbons.
2. These chemicals accumulate in a small reservoir adjacent to the gland and are released onto the body surface only as needed. The larvae of certain swallowtail butterflies have eversible glands, called osmeteria, located just behind the head. When a caterpillar is disturbed, it rears up, everts the osmeteria to release a repellent volatile, and waves its body back and forth to ward off intruders.
3. **Induce cleaning** – irritant compounds often induce cleaning behavior by a predator, giving the prey time to escape. Some blister beetles (family Meloidae) produce cantharidin, a strong irritant and blistering agent that circulates in their hemolymph. Droplets of this blood ooze from the beetle's leg joints when it is disturbed or threatened – an adaptation known as reflex bleeding. Irritant sprays are produced by some termites, cockroaches, earwigs, stick insects, and beetles. The notorious bombardier beetles store chemical precursors for an explosive reaction mixture in specialized glands. When threatened, these precursors are mixed together to produce a forceful discharge of boiling hot quinone and water vapor (steam).
4. **Adhesion** – sticky compounds that harden like glue to incapacitate an attacker. Several species of cockroach guard their backsides with a slimy anal secretion that quickly cripples any worker ants that launch an attack. Similarly, members of the soldier caste in nasute termites have nozzle-like heads equipped with a defensive gland that can shoot a cocktail of defensive chemicals at intruders. The compounds, which are both irritating and immobilizing, have been shown to be highly effective against ants, spiders, centipedes, and other predatory arthropods,
5. **Cause pain or discomfort** – Saddleback caterpillars, larvae of the io moth, and various other Lepidopteran larvae have hollow body hairs that contain a painful irritant. Simply brushing against these urticating hairs will cause them to break and release their contents onto your skin. The consequence is an intense burning sensation that may last for several hours. Many ants, bees, and wasps (the aculeate Hymenoptera) delivervenom to their enemies by means of a formidable stinger (modified ovipositor). The venom is a complex mixture of proteins and amino acids that not only induce intense pain but may also trigger an allergic reaction in the victim.

Protective Coloration

Biologists recognize that there is usually an underlying rationale for the great diversity of shapes and colors found in the insect world. We may not know why a particular species has parallel ridges on the pronotum or black spots on the wings, but we can be reasonably certain that this shape or color has contributed in some way, however small, to the overall fitness of the species. It is obvious that at least some of the colors and patterns serve a defensive function by offering a degree of protection from predators and parasites. These patterns, collectively known as protective coloration, fall into four broad categories:

1. **Crypsis**: Insects that blend in with their surroundings often manage to escape detection by predators and parasites. This tactic, called cryptic coloration, involves not only matching the colors of the background but also disrupting the outline of the body, eliminating reflective highlights from smooth body surfaces, and avoiding sudden movements that might betray location. Obviously, this tactic loses much of its effectiveness if an insect moves from one type of habitat to another.
2. **Well-camouflaged** insects usually stay close to home or make only short trips and return quickly to the shelter of their protective cover. Many ground-dwelling grasshoppers and katydids, for example, have colors of mottled gray and brown that helps them "disappear" against a background of dried leaves or gravel. On the other hand, closely related species that live in foliage are usually a shade of green that matches the surrounding leaves. The larvae of some lacewings improve their camouflage by attaching bits of moss or lichen from their environment onto the dorsal side of their body.
3. **Mimesis**: Some insects "hide in plain sight" by resembling other objects in the environment. A thorn could really be a treehopper; a small twig might be a walkingstick, an assassin bug, or the caterpillar of a geometrid moth; and sometimes a dead leaf turns out to be a katydid, a moth, or even a butterfly. This "mimicry" of natural objects is often known as mimesis. It goes far beyond imitation of plant parts:

   Some swallowtail larvae resemble bird droppings; others have false eyespots on the thorax that create a convincing imitation of a snake's head.

   The likeness of a caterpillar can be found on the outer edge of many lepidopteran wings, perhaps serving to fool predatory birds that may peck at the wing margin instead of the butterfly's body.

   Many butterflies and moths have eyespots on the wings that emulate the face of an owl or some other large animal.

   Slug caterpillars and hag moth larvae look like hair balls or small furry mammals.
4. **Warning Colors**
5. Insects that have an active means of defense (like a sting or a **repellent spray**) frequently display bright colors or contrasting patterns that tend to attract attention. These visually conspicuous insects illustrate aposematic coloration, a term derived from the Greek words *apo-* (from a distance) and *sema* (a sign or signal) – meaning "a signal from afar". A predator quickly learns to associate the distinctive coloration with an "unpleasant" outcome, and one such encounter is usually enough to insure avoidance of that prey in the future. A few individuals will die as sacrifices, but for the species as a whole, it pays to advertise!

6. **Mimicry**:If a distinctive visual appearance is sufficient to protect an unpalatable insect from predation, then it stands to reason that other insects might also avoid predation by adopting a similar appearance. This ploy, essentially a form of "false advertising", was first recognized and described by Henry W. Bates in 1861. Today, it is commonly known as Batesian mimicry. Viceroy butterflies (mostly palatable to birds) are largely protected from predation because they resemble monarch butterflies (very distasteful).

   Many species of bee flies, flower flies, robber flies, and clear-winged moths are similarly protected because they mimic the appearance (and often the behavior) of stinging bees and wasps. Batesian mimicry is usually a successful strategy as long as the model and mimic are found in the same location, the mimic's population size is smaller than that of the model, and predators associate the model's appearance with an unpleasant effect

   In 1879, Fritz Müller recognized that two or more distasteful species often share the same aposematic color patterns. Many species of wasps, for example, have alternating bands of black and yellow on the abdomen. This defensive tactic, commonly known as Müllerian mimicry, benefits all members of the group because it spreads the liability for "educating the predator" over more than one species. In fact, as the number of species in a Müllerian complex increases, there is a greater selective advantage for each individual species.

   Mimicry has been carried to extremes in some tropical Lepidoptera where both related and unrelated species resemble each other in size, shape, color, and wing pattern. Collectively, these butterflies (and sometimes moths) form mimicry rings that may include both palatable and unpalatable species. In South America, for example, longwing butterflies (Family Nymphalidae) form a mimicry ring that includes at least twelve different species (including one moth).

   Although natural selection favors individuals in a population with the best camouflage or mimicry, it also favors the predator or parasite with the best prey-finding acumen. As a result of these competing interests, coevolution between predator and prey populations inevitably leads to an ongoing escalation of offensive and defensive measures – a scenario that Leigh Van Valen of Chicago University describes as an evolutionary "arms race".

   In order to survive in the arms race, both predator and prey must constantly evolve in response to the other's changes. Failure to "keep up" concedes a competitive advantage to the opponent and may lead to extinction. The idea that perpetual change is necessary just to maintain the status quo has been coined the Red Queen's Hypothesis. This name refers to a scene from the stories of Alice in Wonderland by Lewis Carroll. In *Through the Looking Glass*, Alice meets a chess piece, the Red Queen. After running hard to follow the Queen, Alice discovers that she has not moved from where she started. Asked about this paradox, the Red Queen replies, "Here, you see, it takes all the running you can do to keep in the same place."

Concept of Predators, Satiation

Predator satiation (less commonly called predator saturation) is an **antipredator adaptation** in which prey occur at high population densities, reducing the probability of an individual organism being eaten. When predators are flooded with potential prey, they can consume only a certain amount, so by occurring at high densities prey benefit from a safety in numbers effect. This strategy has evolved in a diverse range of prey, including notably many species of plants, insects, and fish. Predator satiation can be considered a type of refuge from predators.

As available food increases, a predator has more chances of survival, growth, and reproduction. However, as food supply begins to overwhelm the predator's ability to consume and process it, consumption levels off. This pattern is evident in the functional response of type II. There are also limits to population growth (numerical response), dependent on the generation time of the predator species.

This phenomenon is particularly conspicuous when it takes the form of masting, the production of large numbers of seeds by a population of plants.

Some periodical cicada (*Magicicada*) species erupt in large numbers from their larval stage at intervals in years that are prime numbers. At high density sites, research finds that the number eaten by birds does not increase with the number of cicada individuals and the risk of predation for each individual decreases.

In contrast to predator satiation, a different pattern is seen in response to mutualistic consumers, which benefit an organism by feeding from it (such as frugivores, which disperse seeds). For example, a vine's berries may ripen at different times, ensuring frugivores are not swamped with food and so resulting in a larger proportion of its seeds being dispersed.

Evolution of Life History Strategies

The analysis of life histories is essentially aiming for an understanding of the evolution of the temporal sequence of decisions between behavioural alternatives during an organism's entire life-time. Conspicuous behavioural characteristics of an organism's life history are transitions between activity and inactivity, such as between diapauses and development, between sleeping and being awake, and plainest between life and death. Certainly choices between other alternatives are of interest, too: the transition from growth to reproduction, the number and timing of reproductive events, the number of offspring, or the choice of a suitable habitat for foraging or oviposition.

These and many other characteristics of life vary both between species and betweenindividuals of the same species and their evolution has been addressed in numerable life history studies (Roff, 1992; Stearns, 1992).

Insect life histories and mating strategies

1. Life history evolution and polymorphism in seasonal environments
2. Reproductive strategies and sexual conflict
3. Host use and life-histories of lichen-feeding moths

1. Life History Evolution and Polymorphism in Seasonal Environments

It focus on how insect life history strategies vary along a climatic cline. In insects, the number of generations emerging per season decreases with increasing latitude. As the favourable time for reproduction and juvenile development decreases, the selection pressure on life-cycles to fit the time-constraints becomes intense. It explore variation in reproductive strategies and life-history traits both within and between populations.

What maintains colour morphs in butterflies?

Variability between individuals is the key material for evolutionary changes but factors that maintain variation within a population are often poorly understood. A widely distributed butterfly *Pieris napi* (Lepidoptera; Pieridae) has distinct colour morphs in Finland. These morphs partly occur sympatric in some areas. We assess the roles of sexual and natural selection on colour polymorphism in this butterfly.

2. Reproductive Strategies and Sexual Conflict

Female polyandry is affected by multiple costs and benefits of mating. Female and male interests over mating decisions are often in conflict, but there are mutual in matings as well. We explore variation in mating strategies and sexual conflict between generations and among species.

Evolution of mating plugs in butterflies: sexual conflict or mutual benefits?

Mating plugs have been interpreted as male defensive adaptations against sperm competition with sexually antagonistic implications, but empirical evidence is scarce or inadequate so far. Sperm competition in particular may favour male characteristics that have malign influence on female fitness creating apparent short term conflict of interests between mating pair. This is, however, overly simplistic scenario if indirect benefits for a female are not taken into consideration. We study the fitness consequences of male derived mating plug in the Clouded Apollo butterfly *Parnassius mnemosyne* (Lepidoptera; Papilionidae).

Sexual conflict and polyandry in waterstriders

Females of most water strider species (Heteroptera; Gerridae) mate repeatedly. Multiple matings have been explained as convenience polyandry. However, possible benefits deriving from repeated matings to females have rarely been investigated within this group of insects. We explore both genetic and material benefits of repeated matings using the waterstrider *Aquarius paludum* as a model organism.

3. Host Use and Life-Histories of Lichen-Feeding Moths

The project is aimed at better understanding on the evolution of host selection and adaptive plasticity in insect life histories. The project will also consider optimal growth strategies of direct breeders and over-wintering individuals. The experiments are conducted using lichen-feeding species, *Cleorodes lichenaria* (Lepidoptera; Geometridae) and *Eilema* spp. (Lepidoptera; Arctiidae). Research questions are divided into four specific topics:

- ☆ The role of genetic and environmental factors in the production of a facultative second generation.
- ☆ Physiological and ecological costs of high growth rates.
- ☆ Implications of juvenile host use on larval colouration with a special reference to the role of sequestered lichen secondary chemicals on the larval defence.
- ☆ Mating systems and resource allocation in reproduction, and consequent costs and benefits for females and their offspring.

It thus seems that the evolutionary transition isless problematic than theory suggests. Considerable efforthas been put into identifying factors that facilitate the spread of a tolerant allele. There is now theoretical and empirical evidence that costs of ghting, female-biased sex ratios, size asymmetries among sibs, single-sex broods, and, especially, reduced larval mobility all facilitate the spread of a tolerant allele. The evolution of gregariousness critically depends on the degree of interactions between larvae; immobility of larvae facilitates gregariousness, while superparasitism, that is, the parasitization by more than one conspecic female, operates against the spread of a tolerant allele

The above-mentioned models use a genetic approach, mainly focusing on the interaction between the offspring of a single female. In this article, we envisage an additional approach, where we focus on ecological conditions that inuence the strength of interaction between the offspring of different (tolerant and nontolerant) females. We expect behavioral responses to spatial heterogeneity (*e.g.*, searching in a host density-[in]dependent manner) to affect the probability of interaction between solitary and gregarious parasitoids that are simultaneously exploiting a host population. Such an ecological approach has not previously been considered. Similarly, differences in trade-offs between solitary and gregarious life-history strategies, which we believe are essential for understanding the ecology and evolution of these different strategies, have scarcely been considered. Solitary parasitoids produce only one offspring per host irrespective of the amount of resources a host provides. Gregarious parasitoids, by contrast, can produce more offspring per host, even without severe scramble-type competition. This especially holds for hemolymph feeding parasitoids that allow their hosts to develop further after. While a solitary larva leaves a considerable portion of the host unused, gregarious development can make full use of the host resource.

General life-history theory predicts a trade-off between fecundity and longevity (a cost of reproduction . Because of this trade-off, gregarious parasitoids that use more eggs per host may have shorter life spans than solitary parasitoids), which translates to fewer opportunities for encountering hosts. Thus, there is a need to consider evolution of gregarious parasitoid life history in a spatial context.

Here we use deterministic models to detect conditions that permit the evolution of the gregarious life history in solitary parasitoid populations. Where analytical solutionsare unfeasible, we use numerical solutions.

The Model

In the case of larval tolerance, competition between solitary and gregarious forms is highly asymmetric. Hence, the ûtness of a gregarious individual depends on the frequency of interactions with solitary individuals (but not the other way around). We therefore apply a game theoretical approach; that is, we search for conditions under which a gregarious mutant is able to invade a population of solitary parasitoids. We are not concerned about the genetic system involved. Rather we assume that strategies replicate asexually to keep the ecological interactions we are interested in tractable. We realize that haplodiploid inheritance complicates the spread of a gregarious allele considerably as a result of the fact that some of the offspring will be of mixed phenotype and gregarious offspring are lost through within-brood interactions. This problem has been dealt with earlier, and we focus entirely on the interactions between broods. Inclusion of mixed phenotype broods would make the conditions for the spread of gregariousness much more stringent than stated in our results. Nevertheless, the results of our models elucidate situations that in general will facilitate or hamper the evolution of gregariousness. In our approach, we link interactions between solitary and gregarious parasitoids to their foraging behavior in patchy habitats. We calculate the marginal clutch size, that is, a mutant's clutch size that leads to the same number of surviving offspring as an average solitary parasitoid produces. Any higher clutch size of the gregarious parasitoid would then allow the gregarious strategy to spread in a population of solitary parasitoids. Clearly, the marginal clutch size will depend on the reproductive success a solitary parasitoid can gain in the habitat and the chance that the gregarious mutant will be the sole exploiter of a host wherein her offspring escape fatal competition with solitary offspring.

Chapter 13

Community Ecology: Concept of Guild, Organisation of Community

Community Ecology

Theory is equivocal about invasion success, implicating various combinations of r, a population's intrinsic rate of increase, K, its 'carrying capacity', enemy-imposed death rates and the coefficient of variation in population numbers as determinants of establishment. Data for a wide variety of organisms, both vertebrates and invertebrates (including insects), accidentally or deliberately introduced by man into the British Isles, appear to show that the probability of establishment of an invader is positively correlated with body size. These data are consistent with the idea that the amplitude of population fluctuations is the main determinant of invasion success, but not with theoretically expected effects of r or K (because populations of small organisms generally fluctuate more than populations of large organisms, but both r and K decrease with increasing body size). However, data for various insect orders introduced into Britain show exactly the opposite trend, with probability of establishment decreasing with increasing body size, and hence possibly with decreasing r and/or K. Possible reasons for these contradictory results, including biases in the data, are discussed. Finally, data from a variety of sources, including insects released as biological control agents, show that enemies (parasitoids and predators) are often a cause of failure to establish particular species of introduced insects. These data add further to the uncertainty about the main determinants of invasion ability. The only clear conclusion to emerge from a combination of theoretical and empirical studies is that r alone is not obviously the main, or even an important determinant of invasion success.

Concept of Guild

A guild (or ecological guild) is any group of species that exploit the same resources, often in related ways. It does not follow that the species within a guild occupy the same, or even similar, **ecological niches**. Guilds are defined according to the locations, the attributes, and the activities of their component species; for example, their mode of acquiring nutrients, their mobility, and the zones of their habitat that they occupy or otherwise exploit. The number of guilds occupying an ecosystem is termed its *disparity*. Members of a guild within a given ecology could be competing for some resources (such as space or light), while cooperating in resisting wind stresses, attracting pollinators, or detecting predators, such as happens among savannah-dwelling antelope and zebra.

A guild does not typically have strict, or even clearly defined boundaries. A broadly-defined guild will practically always have constituent guilds; for example, grazing guilds will have some species that concentrate on coarse, plentiful forage, while others concentrate on low-growing, finer plants. Each of those two sub-guilds may be regarded as guilds in appropriate contexts, and they might in turn have sub-guilds in more closely selective contexts. Some authorities even speak of a fractal resource model. This is a concept that arises in several related contexts, such as the metabolic theory of ecology, the scaling pattern of occupancy, and spatial analysis in ecology, all of which are concepts fundamental in defining guilds.

An ecological guild is not to be confused with a taxocene, a group of phylogenetically related organisms in a community that do not necessarily share the same or similar niches. Nor is a guild the same as a trophic species.

Browse

The plant material eaten is known as *browse* and is naturally taken straight from the plant, though owners of livestock such as goats and deer may cut twigs or branches for feeding to the stock. In temperate regions, owners take browse before leaf fall, then dry and store it as a winter feed supplement. In time of drought, herdsmen may cut branches from beyond the reach of their stock, as forage at ground level. In the tropical regions, where population pressure leads owners to resort to this more often, there is a danger of permanent depletion of the supply. Animals in captivity may be fed browse as a replacement for their wild food sources; in the case of pandas, the browse may consist of bunches of banana leaves, bamboo shoots, slender pine, spruce, fir and willow branches, straw and native grasses.

Folivore

In zoology, a folivore is a **herbivore** that specializes in eating leaves. Mature leaves contain a high proportion of hard-to-digest cellulose, less energy than other types of foods, and often toxic compounds. For this reason folivorous animals tend to have long digestive tracts and slow metabolisms. Many enlist the help of symbiotic bacteria to release the nutrients in their diet. Additionally, as has been observed in folivorous primates, animals may exhibit a strong preference for immature leaves, which tend to be higher in energy and protein and lower in fibre and poisons than more mature leaves.

Grazers

Grazing has two distinct meanings. First, grazing is a method of feeding in which a herbivore feeds on plants such as grasses, or other multicellular organisms such as algae. Many small selective **herbivores** follow larger grazers, who skim off the highest, tough growth of plants, exposing tender shoots. For terrestrial animals, grazing is normally distinguished from ***browsing*** in that grazing is eating grass or forbs, and browsing is eating woody twigs and leaves from trees and shrubs. Grazing differs from true predation because the organism being grazed upon is not generally killed. Grazing differs from parasitism as the two organisms do not live together, nor is the grazernecessarily so limited in what it can eat . Water animals that feed for example on algae found on stones are called grazers-scrapers. Grazers-scrapers feed also on microorganisms and dead organic matter on various substrates.

Second, grazing is a method of agriculture in which domestic livestock are used to convert grass and other forage into meat, milk and other products.

Organisation of Community

The extent to which insect communities are organized is poorly understood because few studies have examined both population–level processes and community–level patterns. Furthermore, our view of phytophagous insect populations is probably biased by the more frequent attention given to economic pests whose dynamics might be expected to be somewhat exceptional. Herein, we report on the population dynamics and community–level features in a diverse, native fauna of phytophagous insects associated with goldenrod (Solidago altissima), a native perennial plant that forms a dominant, long–persisting element in old–field succession. The data consist of censuses taken in six consecutive years at 16 old fields in the Finger Lakes Region of New York; five additional stands were followed for four consecutive years. Our analyses address two questions: (a) to what extent is the functional structure of the community (as reflected in such attributes to the herbivore load, guild spectrum, and dominance hierarchy) maintained by compensatory changes in the densities of the member species and (b) are certain traits associated with a species' tendency to fluctuate in density or to dominate the community? In general, populations of goldenrod insects fluctuated less than those of insects reported in the literature. Few were abundant; only 7 of the 138 species in the goldenrod fauna ever reached densities at which their biomass exceeded 0.1 per cent of the leaf biomass. The functional structure of the community, as reflected in the total herbivore load and the proportions of that load contributed by various guilds, varied widely in space and time. A small subset of the fauna remained dominant throughout the study; these species, whose populations occasionally irrupted, retained their high rank in relative abundance even during those periods when their populations were waning. There was no evidence for compensatory changes in the densities of species within guilds, as would be expected if a relatively distinct set of limiting resources were available to insects with different feeding styles. Species were no more likely to exhibit negative correlations with their guild mates than they were with members of other guilds. There was also little evidence of sporadic or diffuse competition within the fauna as a whole; even during population outbreaks, the dominant species rarely engendered decreases in the biomass or diversity of the remaining species, and increases in the total herbivore load were seldom associated with losses of species from the community. Several species were positively associated with one another on the basis of similar habitat requirements or the use by one species of conditions created by the presence of another. The community was predictable only in the sense that dominant species remained dominant and rare species rare. The abundance and population variability of species were not correlated with such traits as body size, generation time, or host range. Taxon–related trends, on the other hand, stood out; coleopterans tended to be abundant whereas lepidopterans were rare, and hemipterans tended to fluctuate widely whereas dipterans remained relatively constant. Furthermore, species that were clumped in space (onto stems within fields) tended to be abundant and variable. The "boom and bust" dynamics of these aggregating species may be related to the ability of colonies to survive the inroads of generalist predators. The low degree of regional synchrony in the performance of populations at different sites suggests that large–scale, weather–driven fluctuations are uncommon. Our analysis of this well–developed, native insect fauna reveals a community that has a somewhat predictable structure that reflects the idiosyncracies of the component species. These species do not fluctuate in an integrated community–structuring manner, but as an outcome of each species performing at its "individualistic" level, the community displays a characteristic dominance hierarchy. The members of this diverse fauna rarely achieve densities at which they compete with one another. Our results are consistent with Hairston, Smith, and Slobodkin's hypothesis (1960) which predicts that terrestrial herbivores rarely deplete plant resources to levels at which there is widespread interspecific competition.

Chapter 14

Relative Distribution of Organism, Concept of Diversity: The Wallacian View

Relative Distribution of Organism

Relative species abundance is a component of biodiversity and refers to how common or rare a species is relative to other species in a defined location or community. Relative species abundances tend to conform to specific patterns that are among the best-known and most-studied patterns in macroecology.

Relative species abundance and species richness describe key elements of biodiversity. Relative species abundance refers to how common or rare a species is relative to other species in a given location or community.

Usually relative species abundances are described for a single **trophic level**. Because such species occupy the same trophic level they will potentially or actually compete for similar resources. For example, relative species abundances might describe all terrestrial birds in a forest community or all **planktonic copepods** in a particular marine environment.

Relative species abundances follow very similar patterns over a wide range of ecological communities. When plotted as a histogram of the number of species represented by 1, 2, 3, ...,*n* individuals usually fit a hollow curve, such that most species are rare, (represented by a single individual in a community sample) and relatively few species are abundant (represented by a large number of individuals in a community sample) (Figure 1). This pattern has been long-recognized and can be broadly summarized with the statement that "most species are rare". For example, **Charles Darwin** noted in 1859 in the **Origin of Species** that "*...rarity is the attribute of vast numbers of species in all classes...*"

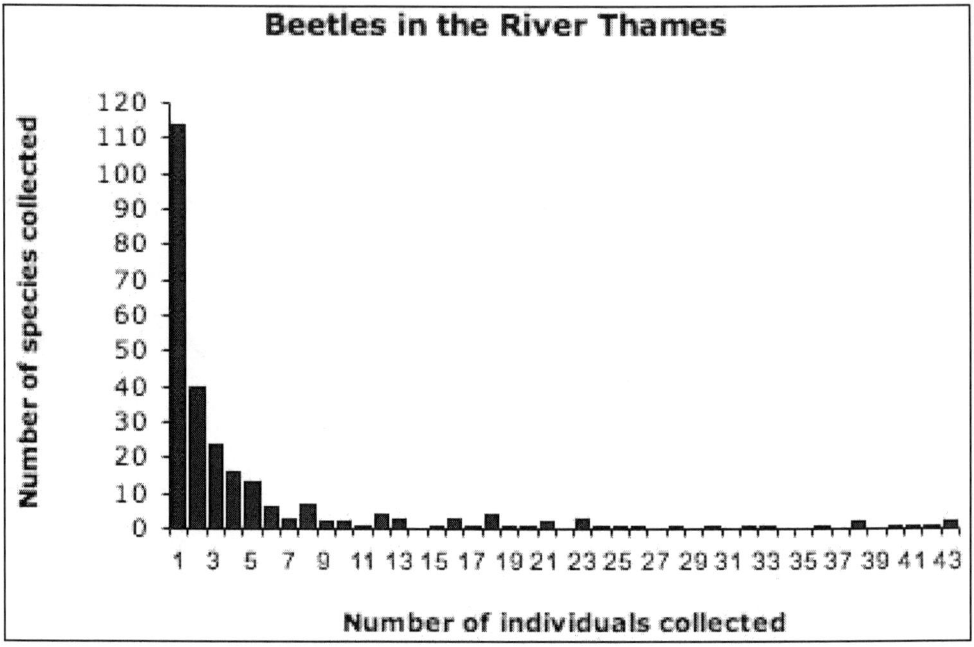

Figure 1: Relative Species Abundance of Beetles Sampled from the River Thames Showing the Universal "Hollow Curve". (Derived from data presented in Magurran (2004) and collected by C.B. Williams (1964).

Species abundance patterns can be best visualized in the form of relative abundance distribution plots. The consistency of relative species abundance patterns suggests that some common macroecological "rule" or process determines the distribution of individuals among species within a trophic level.

Distribution Plots

Relative species abundance distributions are usually graphed as frequency histograms ("Preston Plots"; Figure 2) or rank-abundance diagrams ("Whittaker Plots"; Figure 3).

Frequency histogram (Preston Plot):

x-axis: logarithm of abundance bins (usually $\log_2$ (because this was historically a simple way to approximate the natural log))

y-axis: number of species at given abundance

Rank-abundance diagram (Whittaker Plot):

x-axis: species list, ranked in order of descending abundance (*i.e.* from common to rare)

y-axis: logarithm of per cent relative abundance

When plotted in these ways, relative species abundances from wildly different data sets show similar patterns: frequency histograms tend to be right-skewed (*e.g.*

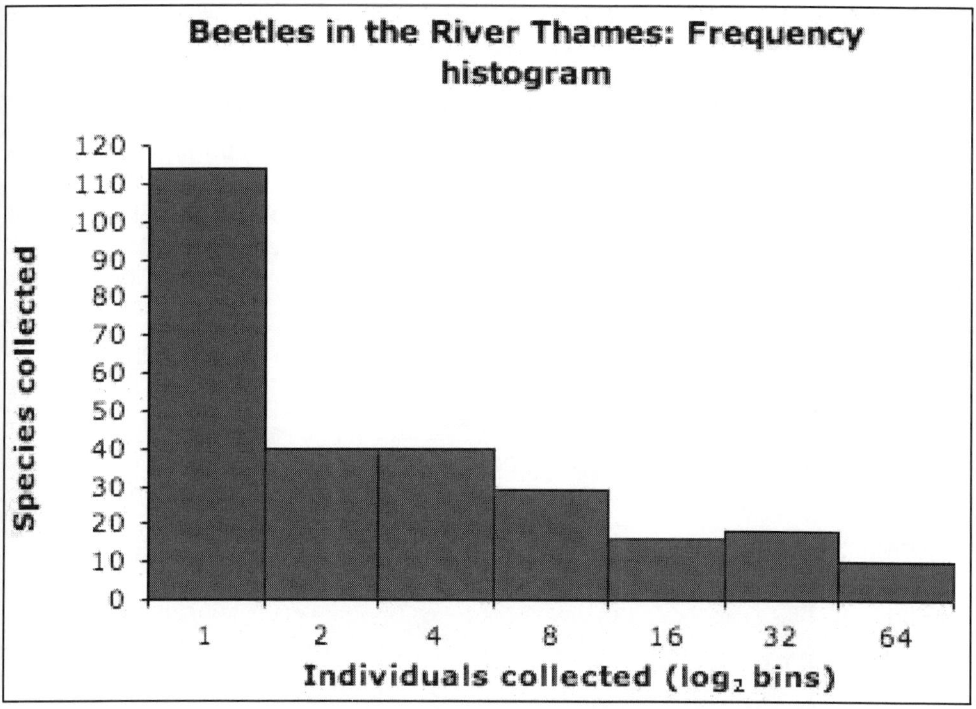

Figure 2: ***Preston Plot*** **of Beetles Sampled from the River Thames Showing a Strong Right-Skew.**

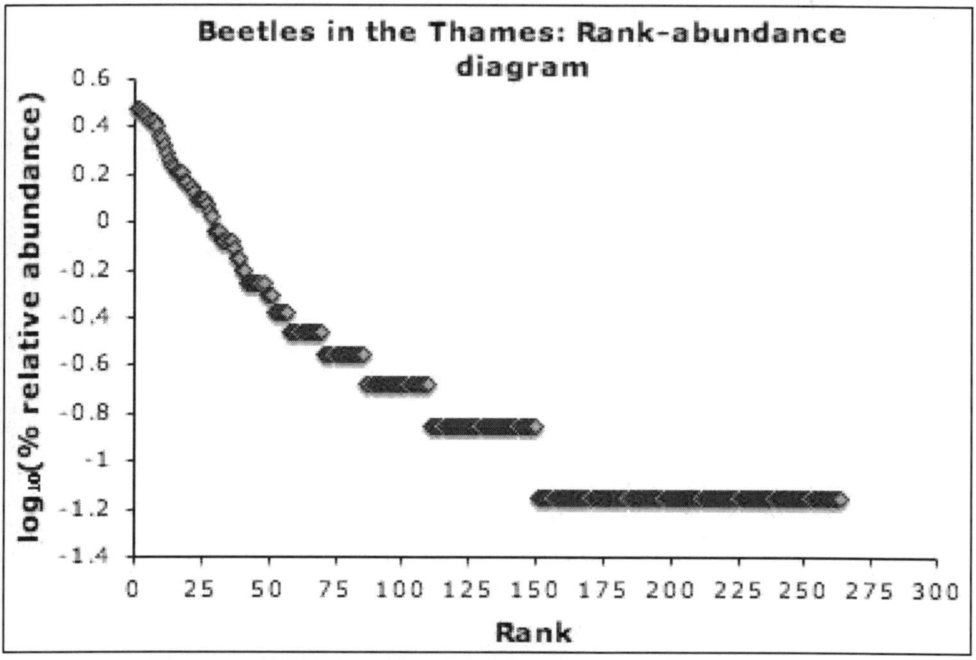

Figure 3: ***Whittaker Plot*** **of Beetles Sampled from the River Thames Showing a Slight "s"-Shape.**

Figure 2) and rank-abundance diagrams tend to conform to the curves illustrated in Figure 4.

Understanding Relative Species Abundance Patterns

Researchers attempting to understand relative species abundance patterns usually approach them in a descriptive or mechanistic way. Using a descriptive approach biologists attempt to fit a mathematical model to real data sets and infer the underlying biological principles at work from the model parameters. By contrast, mechanistic approaches create a mathematical model based on biological principles and then test how well these models fit real data set

Descriptive Approaches

Geometric Series (Motomura 1932)

I. Motomura developed the geometric series model based on benthic community data in a lake. Within the geometric series each species' level of abundance is a

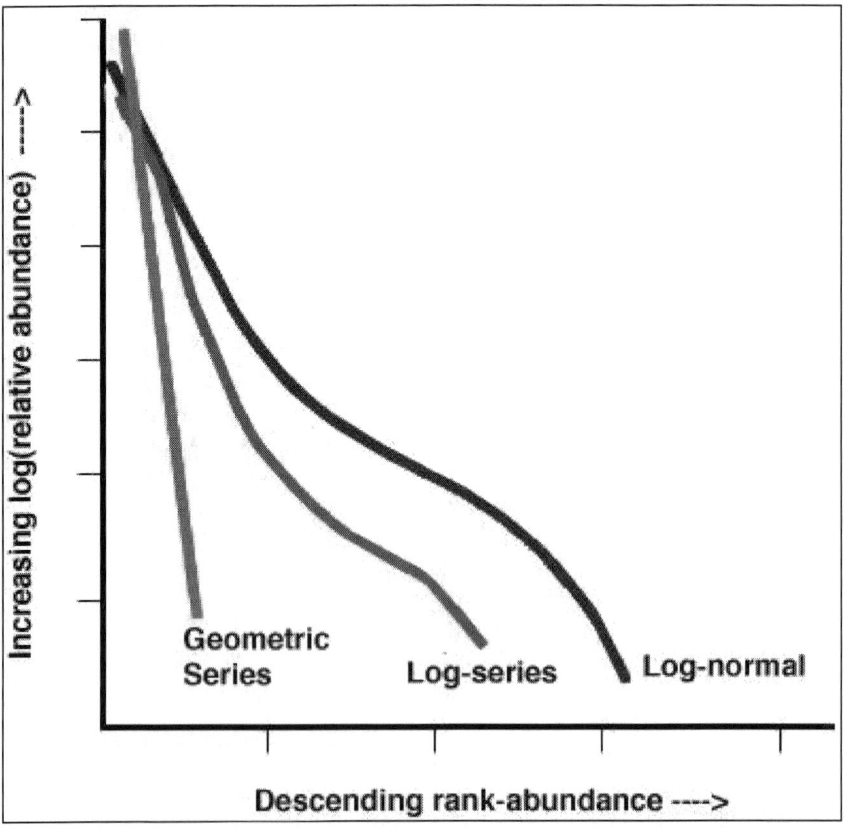

Figure 4: Generic Rank-Abundance Diagram of Three Common Mathematical Models Used to Fit Species Abundance Distributions: Motomura's Geometric Series, Fisher's Log Series, and Preston's Log-Normal Series (Modified from Magurran 1988).

sequential, constant proportion (*k*) of the total number of individuals in the community. Thus if *k* is 0.5, the most common species would represent half of individuals in the community (50 per cent), the second most common species would represent half of the remaining half (25 per cent), the third, half of the remaining quarter (12.5 per cent) and so forth.

Although Motomura originally developed the model as a statistical (descriptive) means to plot observed abundances, the "discovery" of his paper by Western researchers in 1965 led to the model being used as a **niche apportionment model** – the "niche-preemption model". In a mechanistic model *k* represents the proportion of the resource base acquired by a given species.

The geometric series rank-abundance diagram is linear with a slope of –*k*, and reflects a rapid decrease in species abundances by rank (Figure 4). The geometric series does not explicitly assume that species colonize an area sequentially, however, the model fits the concept of niche preemption, where species sequentially colonize a region and the first species to arrive receives the majority of resources. The geometric series model fits observed species abundances in highly uneven communities with low diversity. This is expected to occur in terrestrial plant communities (as these assemblages often show strong dominance) as well as communities at early successional stages and those in harsh or isolated environments (Figure 5).

Logseries (Fisher et al 1943)

$$S = \alpha \ln \left(1 + \frac{N}{\alpha}\right)$$

where,

$S =$ The number of species in the sampled community

$N =$ the number of individuals sampled

$\alpha =$ a constant derived from the sample data set

The logseries was developed by Ronald Fisher to fit two different abundance data sets: British moth species (collected by Carrington Williams) and Malaya butterflies (collected by **Alexander Steven Corbet).** The logic behind the derivation of the logseries is varied however Fisher proposed that sampled species abundances would follow a negative binomial from which the zero abundance class (species too rare to be sampled) was eliminated. He also assumed that the total number of species in a community was infinite. Together, this produced the **logseries distribution** (Figure 4). The logseries predicts the number of species at different levels of abundance (*n*individuals) with the formula:

$$S_n = \frac{\alpha x^n}{n}$$

where,

S = The number of species with an abundance of *n*

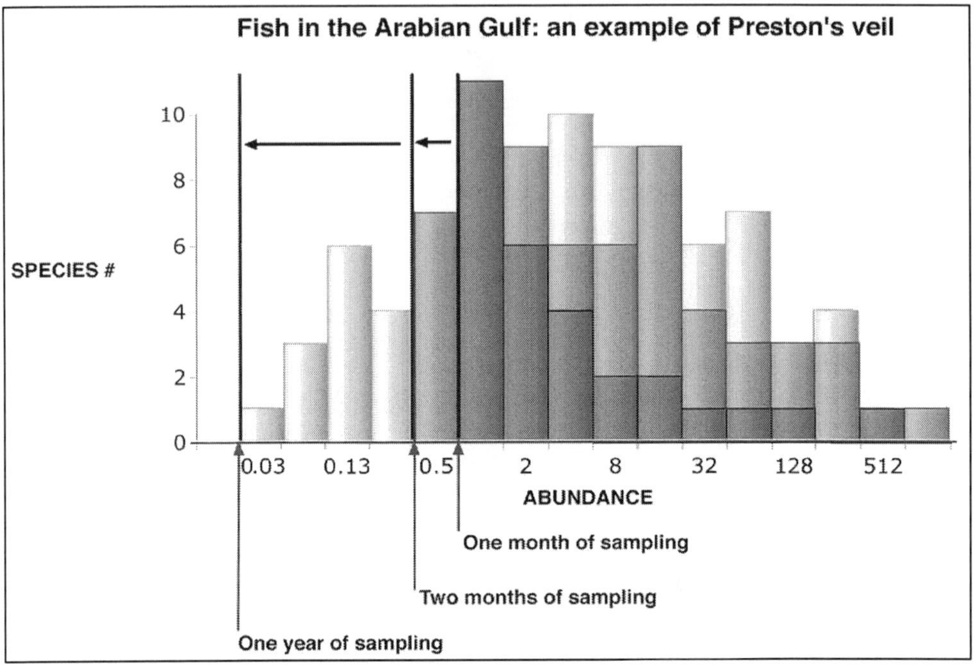

Figure 5: An Example of Preston's Veil. Fish species abundances sampled using repeated trawling over a one-month (blue bars), two month (gold bars) and one-year period (yellow). One year of sampling indicates that the fish community is log-normally distributed (Derived from Magurran 2004).

x = A positive constant ($0 < x < 1$) which is derived from the sample data set and generally approaches 1 in value

The number of species with 1, 2, 3,..., n individuals are therefore:

$$\alpha, \frac{\alpha x^2}{2}, \frac{\alpha x^3}{3}, \ldots, \frac{\alpha x^n}{n}$$

Fisher's Constants

The constants α and x can be estimated through iteration from a given species data set using the values S and N. Fisher's dimensionless α is often used as a measure of biodiversity, and indeed has recently been found to represent the fundamental biodiversity parameter, θ, from neutral theory.

Log Normal (Preston 1948)

Using several data sets (including breeding bird surveys from New York and Pennsylvania and moth collections from Maine, Alberta and Saskatchewan) **Frank W. Preston** (1948) argued that species abundances (when binned logarithmically in a Preston plot) follow a **Normal (Gaussian) distribution,** partly as a result of the **Central Limit Theorem** (Figure 4). This means that the abundance distribution is **Lognormal**. According to his argument, the right-skew observed in species abundance frequency histograms (including those described by Fisher *et al.* (1943) was, in fact, a sampling artifact. Given that species toward the left side of the x-axis are increasingly rare, they may be missed in a random species sample. As the sample size increases however, the likelihood of collecting rare species in a way that accurately represents their abundance also increases, and more of the normal distribution becomes visible. The point at which rare species cease to be sampled has been termed *Preston's veil line*. As the sample size increases Preston's veil is pushed farther to the left and more of the normal curve becomes visible (Figure 5). Interestingly, Williams' moth data, originally used by Fisher to develop the logseries distribution, became increasingly lognormal as more years of sampling were completed.

Calculating Theoretical Species Richness

Preston's theory has an interesting application: if a community is truly lognormal yet under-sampled, the lognormal distribution can be used to estimate the true species richness of a community. Assuming the shape of the total distribution can be confidently predicted from the collected data, the normal curve can be fit via statistical software or by completing the Gaussian formula:

$$n = n_0 e^{-(aR)^2}$$

where,

n_0 is the number of species in the modal bin (the peak of the curve)

n is the number of species in bins R distant from the modal bin

a is a constant derived from the data

It is then possible to predict how many species are in the community by calculating the total area under the curve (N):

$$N = \frac{n_0\sqrt{\pi}}{a}$$

The number of species missing from the data set (the missing area to the left of the veil line) is simply N minus the number of species sampled. Preston did this for two lepidopteran data sets, predicting that, even after 22 years of collection, only 72 per cent and 88 per cent of the species present had been sampled.

Yule model (Nee 2003)

The Yule model is based on a much earlier, **Galton–Watson model** which was used to describe the distribution of species among genera. The Yule model assumes random branching of species trees, with each species (branch tip) having the equivalent probability of giving rise to new species or becoming extinct. As the number of species within a genus, within a clade, has a similar distribution to the number of individuals within a species, within a community (*i.e.* the "hollow curve"), Sean Nee (2003) used the model to describe relative species abundances. In many ways this model is similar to **niche apportionment models**, however, Nee intentionally did not propose a biological mechanism for the model behavior, arguing that any distribution can be produced by a variety of mechanisms.

Mechanistic Approaches: Niche Apportionment

Most mechanistic approaches to species abundance distributions use niche-space, *i.e.* available resources, as the mechanism driving abundances. If species in the same **trophic level** consume the same resources (such as nutrients or sunlight in plant communities, prey in carnivore communities, nesting locations or food in bird communities) and these resources are limited, how the resource "pie" is divided among species determines how many individuals of each species can exist in the community. Species with access to lots of resources will have higher carrying capacities than those with little access. Mutsunori Tokeshi later elaborated niche apportionment theory to include niche filling in unexploited resource space. Thus, a species may survive in the community by carving out a portion of another species' niche (slicing up the pie into smaller pieces) or by moving into a vacant niche (essentially making the pie larger, for example, by being the first to arrive in a newly available location or through the development of a novel trait that allows access previously unavailable resources). Numerous **niche apportionment models** have been developed. Each make different assumptions about how species carve up niche-space.

Unified neutral theory (Hubbell 1979/2001)

The Unified Neutral Theory of Biodiversity and Biogeography (UNTB) is a special form of mechanistic model that takes an entirely different approach to community composition than the niche apportionment models. Instead of species populations reaching equilibrium within a community, the UNTB model is dynamic, allowing for continuing changes in relative species abundances through drift.

A community in the UNTB model can be best visualized as a grid with a certain number of spaces, each occupied with individuals of different species. The model is **zero-sum** as there are a limited number of spaces that can be occupied: an increase in the number of individuals of one species in the grid must result in corresponding decrease in the number of individuals of other species in the grid. The model then uses birth, death, immigration, extinction and speciation to modify community composition over time.

Hubbell's theta

The UNTB model produces a dimensionless "fundamental biodiversity" number, θ, which is derived using the formula:

$$\theta = 2J_m v$$

where,

J_m is the size of the metacommunity (the outside source of immigrants to the local community)

v is the speciation rate in the model

Relative species abundances in the UNTB model follow a zero-sum multinomial distribution. The shape of this distribution is a function of the immigration rate, the size of the sampled community (grid), and θ. When the value of θ is small, the relative species abundance distribution is similar to the geometric series (high dominance). As θ gets larger, the distribution becomes increasingly s-shaped (log-normal) and, as it approaches infinity, the curve becomes flat (the community has infinite diversity and species abundances of one). Finally, when $\theta = 0$ the community described consists of only one species (extreme dominance).

Fisher's alpha and Hubbell's theta – an interesting convergence

An unexpected result of the UNTB is that at very large sample sizes, predicted relative species abundance curves describe the **metacommunity** and become identical to Fisher's logseries. At this point θ also becomes identical to Fisher's α for the equivalent distribution and Fisher's constant x is equal to the ratio of birthrate: deathrate. Thus, the UNTB unintentionally offers a mechanistic explanation of the logseries 50 years after Fisher first developed his descriptive model.

Concept of Diversity – The Wallacian View

Soon after his return from the Malay Archipelago, Alfred Russel Wallace published one of his most significant papers. The paper used butterflies of the family Papilionidae as a model system for testing evolutionary hypotheses, and included a revision of the Papilionidae of the region, as well as the description of some 20 new species. Wallace argued that the Papilionidae were the most advanced butterflies, against some of his colleagues such as Bates and Trimen who had claimed that the Nymphalidae were more advanced because of their possession of vestigial forelegs. In a very important section, Wallace laid out what is perhaps the clearest Darwinist definition of the differences between species, geographic subspecies, and local 'varieties.' He also discussed the relationship of these taxonomic categories to what is now termed 'reproductive isolation.' While accepting reproductive isolation as a cause of species, he rejected it as a definition. Instead, species were recognized as forms that overlap spatially and lack intermediates. However, this morphological distinctness argument breaks down for discrete polymorphisms, and Wallace clearly emphasised the conspecificity of non-mimetic males and female Batesian mimetic morphs in Papilio polytes, and also in P. Memnon, on the grounds of reproductive continuity. Finally, Wallace detailed how natural selection explains various forms of parallel evolution, including mimicry.

Chapter 15

Assessment of Diversity, Diversity Stability, Debate Relevance to Pest Management

Assessment of Diversity

Necessary steps in a proper study of insect biodiversity

- ☆ Definition of objectives
- ☆ Gathering of existing and background information
- ☆ Development of a plan for the project as a whole
- ☆ Definition of level of detail
- ☆ Site selection
- ☆ Taxon selection
- ☆ Duration of study
- ☆ Selection of sampling methods
- ☆ Quality control of actual sampling
- ☆ Sorting and preparation of samples
- ☆ Identification of material
- ☆ Data management
- ☆ Curation and the disposition of specimens
- ☆ Publication and dissemination of information

Insects thus have great potential for understanding ecosystems and as measures of ecosystem health, but the incompleteness of knowledge and the limitation of resources increase the difficulty of work on insect biodiversity. Therefore, careful targeting of any study is essential.

Definition of Objectives

Any valid scientific study has clearly defined objectives. Therefore, studies of biodiversity should attempt to answer specific questions, and not simply generate isolated lists of species. At the same time, studies of biodiversity are more valuable if they are orientated in a wide spatial, temporal, and social context. In particular, biodiversity studies aim to establish a baseline to assess differences from place to place, under different regimens, or from the present to the future. Such projected comparisons require that procedures be standardized as far as possible, that findings are placed in the context of information available previously, that material be retained for future use, and that information be validly.

Long-term use of the data therefore, is one important objective, but it cannot be the only one. Answers to apposite shorter term questions also will drive the design of the study, and these questions have to be asked in an cological setting (cf. Lehmkuhl *et al.*, 1984): the aim of a study is to find out in some relevant respect what is going on, even though the basic work still is the collection, accumulation and maintenance of large samples of material. For example, valid specific objectives would include the impact of change (artificial disturbance, pollution, etc.) on ecosystem function or persistence, as estimated by the diversity of key trophic groups.

Knowledge of Background Information

Existing published and unpublished information helps in developing specific study design, as well as for comparison and verification of results. Background knowledge allows gaps to be identified and studies having a specific purpose to be designed (Rosenberg *et al.*, 1979). Some effort will be required to do this, because background information about biodiversity is scattered in a variety of taxonomic, ecological and geographical groupings. Directly relevant information or previously collected specimens may be available, and using this material may be more cost effective than repeating a full sampling program provided that the earlier work was carried out adequately.

By the same token, especially when relatively little data exist already, it is important as soon as possible to summarize in a tidy way ongoing findings as they accumulate, to help capitalize on the core work in subsequent seasons. For example, a valuable baseline of the diversity of old-growth Douglas-fir forests in Oregon was published to bring together the findings of various specialists (Parsons *et al.*, 1991).

The Overall Plan

The key to an effective overall plan is ensuring that the required funds and personnel will be available over a time frame that is of adequate duration. In particular, resources and expertise for later aspects of a study, such as identification, curation, and publication, must be given deliberate attention. Otherwise, most resources may have been expended (on sampling and sorting, for example) before the later phases, so essential to the completion and wider value of the project, are reached. Many major projects surveying insect diversity in the past lacked adequate follow-through to completion (Rosenberg *et al.*, 1979).

A detailed plan ensures that a project does not overreach itself in early stages, for example by taking too many samples or attempting to inventory too many groups. Of course, sufficient flexibility must be retained to adjust the program in the light of preliminary findings (Lehmkuhl *et al.*, 1984). A specific scientific focus to the plan ensures that usable results can be obtained, and establishment of specific protocols ensures that samples will be adequate, and that specimens will be of a quality that makes them identifiable. For example, insects live in many different microhabitats.

These microhabitat differences, and variations in life-cycles (some stages are more difficult to collect), make the species differentially available for sampling. Consequently, any "generalist" assessment of the biodiversity of an area is likely to miss many species. Such potential deficiencies reinforce the need for detailed attention to sampling protocols in relation to study objectives and target groups.

The need for correct early decisions of this sort means that both systematists and ecologists should be involved in the design of projects, or at least that expert review of general plans for biodiversity sampling take place early in project development, to avoid unpleasant later surprises as to the utility of the work. Biodiversity studies cannot be allowed to follow the route of many "Environmental Impact Assessments", carried out without adequate scientific planning to meet political ends rather than scientific or real social objectives (Schindler 1976), which consequently generated results of low quality.

Moreover, any reasonably broad biodiversity study has to enlist the early cooperation of systematics experts; otherwise, identification of the groups that hold the key to meeting the project objectives may not be feasible. Early attention should also be given to how the results will be used or analyzed. Such considerations influence what is sampled and how data are recorded. For example, despite some difficulties of interpretation numerical "diversity indices" can give useful shorthand ways of characterizing biodiversity (Samways 1984; Magurran 1988; Cousins 1991). Which index will be used, if any, may influence how abundance data are to be recorded, for example.Successful completion of a project requires resources for all elements of that project. For example, expertise for identification and for solving taxonomic problems in key target groups will often have to be purchased. Indeed, increasingly even government agencies (in Canada and elsewhere) are seeking cost recovery for such expert services as identification. In many cases, the necessary expertise will not even be available for outside purchase, and will then have to be staffed and developed from the beginning as part of the project.

The cost of sampling and preparation of samples in a project that surveys biodiversity adequately is very high. Scudder (1996) analyzed such costs, as shown for a range of techniques. He concluded that employing students at $10 per hour to use only the first 8 methods would cost $24,000 per site per season, and identifying the material to family would double the cost. These substantial costs re-emphasize the necessity for longterm planning for resources.

Identification to Major Families of Taxa (except Acari)

Butterflies collected by an expert, and only voucher specimens processed and identified to species and subspecies by the expert collector. Moths sorted by an expert and only specimens of special interest processed. Similarly, funds to curate material and provide resources for its maintenance in appropriate facilities must be arranged for early in the work. Most museums no longer have the means to house specimens provided by various outside.

In summary, attention to an overall plan for the project as a whole ensures not only that the results will address the project objective, but also that the results will be scientifically valid and so can be built upon for the future (Lehmkuhl *et al.*, 1984; Danks and Ball 1993). Many large-scale studies in the past lacked long-term planning, and so generated only incomplete results that were of no real or lasting benefit.

Degree of Detail

Doing the work properly usually requires identification to species. For nearly all objectives it is better to have specific information on carefully chosen groups than family-level information on many. Species-level information is valuable for two main reasons. First, species are the functioning entities in nature, so that ecosystem interactions can normally best be understood using species-level identifications.

For example, work on caddisflies of the genus Ceraclea (formerly Athripsodes) showed that each species has a different tolerance for changes in conditions caused by industrialization (Resh 1976; Resh and Unzicker 1975). Again, seasonal patterns of larval abundance of Baetis mayflies suggested recovery from insecticide treatment of their river habitat, but in fact the "recovery" resulted from the presence of a second, and temporally separated, distinct species of Baetis (Lehmkuhl 1981). Grouping species in the same insect family therefore is not usually appropriate. Indeed, a moderate sized family of insects, for example chironomid midges, contains species differing greatly in relative size and feeding habits, a diversity similar to that of the whole of the birds, for example. It would not generally be thought reasonable to group data about all birds ("kilograms of birds per hectare"); it is not reasonable to do so for insects, despite some past practices stemming from taxonomic difficulties.

The first reason for specific identifications, therefore, is biological reality and applicability of data, which ensure that the results and analysis will be usable to answer questions of interest. A second reason for species-level work is that species names allow information to be associated with each taxonomic entity for future reference. All biological information is collected together and retrieved on the basis of species' names; because such knowledge can be referenced effectively, it can be used and developed through time. Therefore existing information can be integrated with the new findings. Identifications only at higher taxonomic levels do not uniquely document diversity or make it possible to use the information for detailed comparisons. Attention to details of habitat, etc. is also necessary for adequate documentation.

The need for detailed identification is obvious in many walks of life:

Misidentifying even the strain of a pet dog can lead to difficulties if a pup grows ten times larger than expected. The need for detailed identification has nevertheless been overlooked in some projects on biodiversity because those projects were of such broad scope.

Site Selection

At a general level, choosing accessible sites lowers the cost of sampling. Using discrete habitats that are easily recognized also assists in careful and repeatable sampling. Long-term stability of the sites is especially valuable for repeated sampling over time; long-term stability is increased by formal legal protections (*e.g.* national parks) or association with stable institutions (*e.g.* some field stations). Pre-existing and continuing interest by various agencies not only helps to maintain the stability of effort, but also makes available a wider base of information.

For example on climate, vegetation type, or other data of value, as in the U.S. Long Term Ecological Research Sites (Callaghan 1984) or potentially with components of the Canadian Environmental Monitoring and Assessment Network (Environment Canada 1993).

At a more specific level, site choice depends on the objectives of the study, because few studies of biodiversity have sufficient resources to complete a full regional inventory of species from all habitats in a range of places. If the initial project focus is on habitat (*e.g.* forest types), sites should of course be fully representative of each habitat type of interest. If the primary focus is on taxa with particular ecological properties of interest (*e.g.* herb-feeding bugs), emphasizing habitats favoured by the organisms, as determined by expert advice, increases the efficiency of sampling. In any event, sites and habitats should be characterized through ongoing field notes during sampling. Such reports are very helpful later for interpreting the sample data, because they record site variability and provide other valuable clues. For example, seasonal vegetation development and changes in moisture regimens within a given site affect both the presence and the efficiency of trapping for some species.

Selection of Taxa for Study

There are so many groups of insects in most habitats and their numbers are so large that with ordinary resources it is impossible to study them all simultaneously. The choice of taxa depends on utility and feasibility. Other things being equal, this choice depends on the objectives of the study, because groups of different diversity, habitat specificity, dispersal ability, feeding habit, and so on would be expected to have different values for answering the questions posed by the study.

Unfortunately, questions of feasibility usually modify the theoretically ideal choice. For example, the practicality of sampling and sorting differs among groups. Most soil-dwelling groups are more costly to sample than groups that live on the ground surface. Moreover, groups collected must be identified. An initial difficulty is that knowledge in some groups is inadequate to allow specific identifications, although in some taxa morphological species or morphospecies (*i.e.* "species 1", "species 2", etc., as yet unnamed) can be recognized reliably by experts. A second difficulty is that, even for some groups that are reasonably well known, considerable expertise may be required to distinguish the species. The choice of taxa therefore typically is a compromise between scientific relevance and feasibility. Nevertheless, it is unwise to over-emphasize feasibility, because the taxonomic coverage may then be confined to groups that provide little useful information about the objective. For example, characterizing forest types, ages or treatments by means of taxa that occur almost exclusively in large clearings or in pools, and so tend to be similar among the different forest types, is unlikely to be successful.

Conversely, a study intended to ascertain what changes in biodiversity occur in grazed meadows compared to ungrazed sites, and how the changes can be minimized, might target otherwise feasible groups expected to signify the relevant changes more clearly. These changes might include the occurrence and vigour of specific plants and allied microhabitats (reflected by selected plant bugs, for example), ground exposure (influencing grasshoppers, which deposit their eggs in bare soil, and soil mites that respond to local soil moisture and other soil factors), general habitat structure (potentially reflected by ground beetles, for example), and food-chain as well as habitat effects (*e.g.* through study of predaceous wasps). Changes in such species would also indicate if important linkages with other organisms are likely to be disturbed, and suggest further consequences. An informed biodiversity study of this sort then can suggest whether and how mitigation is to be carried out. It is unwise to restrict studies to only one or two easy-to-identify groups. Not only may such a restriction provide too few data to answer the questions at hand, but it is also inefficient because very little use is made of costly samples containing many other potentially instructive taxa. Such "other taxa" are routinely discarded in smaller biodiversity studies, rather than being preserved or made more widely available. Instead, work should normally be done on a relevant but taxonomically and ecologically diverse subset of the11taxa collected. Predators and herbivores, for example, would be expected to show different patterns and to provide different insights.

Duration of the Study

It is not usually possible to sample insect diversity adequately over too short a time frame, both because the temporal development of populations may make individuals available for capture for a relatively short time, and because of natural variations in the occurrence and abundance of individuals. Most sampling devices or techniques target a single stage of the life cycle, especially the adult. However, some adult insects live for a very short time, and when the population emerges synchronously adults may be present in the field for a week or less. Moreover, different species emerge at different characteristic times of year. The emergence of a species may be early or late in any given season, depending on weather. Its availability for trapping even when present (and so whether it will be captured at any given time) depends on the weather and on its abundance. It turn, abundance depends on population processes governed by natural enemies and many other factors.These variations mean that a given species may or may not be captured in a sampling program that is too short or insufficiently intensive. In particular, short visits to sites are likely to provide only a small random selection of the real diversity. Even if a longer lived and therefore more reliably encountered larval stage is sampled instead, for example, it will not be possible to identify most of these larvae without a costly program of rearing and taxonomic analysis. Larvae may nonetheless provide more useful information than adults for some purposes. Taking advantage of this fact requires especially careful planning to overcome additional difficulties of sampling and deficiences of taxonomic knowledge.

Consequently, a study of biodiversity must be planned over a time frame of adequate duration. Local diversity cannot be characterized unless the core data are more-or-less complete. "Hit and run" techniques have been recommended as the only practical way of sampling tropical diversity (Coddington *et al.*, 1992), but are likely to be inadequate except for superficial assessments. Moreover, most changes in biodiversity of potential interest (colonization, succession, population cycles, etc.) accord with long-term natural events, which can be interpreted only when long-term data have been collected. For example, analysis of ten years of population data on the gall midge Taxomyia taxi and its parasitoids failed to demonstrate density-dependent effects; however, extension to a 24-year run of data revealed such effects (Redfern and Cameron 1993). The major events, chiefly climatically driven, that help to govern many ecosystems occur at intervals of several years, which is longer than the normal cycle for the funding of research.

Selection of Sampling Methods

Some general principles that govern the choice of sampling methods are outlined here: in essence, methods should be multiple, targeted, cost-effective and standardized.

Biodiversity is best assessed through several simultaneous sampling methods. All methods have their strengths and weaknesses, and for most objectives a useful cross section of the fauna will be sampled only by using a number of different techniques .Methods must be appropriate for the targeted taxa and habitats. In addition to typical mass-collecting devices, the sampling set can be supplemented by specific methods for target taxa, as long as the uniformity of the core sampling effort is controlled from one study to another. Also, occasional additional material acquired by supplementary techniques, such as searching streamside vegetation or rearing, generally makes the taxonomic work easier. The most cost-effective techniques are passive or behavioural ones (*e.g.* pan trap, pitfall trap, Malaise trap, flight-intercept trap, Berlese funnel); insects come to the trap or collecting vessel, rather than being pursued by the collector. Such techniques yield huge numbers of specimens, leading to good habitat coverage, but creating difficulties of sorting and selection.It is especially important to standardize methods. Only in this way can information from different sites, regions, and times be compared effectively. Such standardization requires careful attention to the number, size and colour of traps, the mesh size of sieves, etc., as well as the day-to-day execution of the sampling. Some attempts to develop standard protocols for sampling insects in particular environments are already underway.Replicating sampling is also important. The need for replication is often overlooked because of the high cost of handling material from multiple traps or sites. However, for both qualitative analysis (species occurrence) and quantitative analysis (numbers of specimens), replication is always essential to answer key questions of interest, which usually take the form of whether the differences in biodiversity between different places or different times are real, or simply reflect sampling variation.

Execution of Sampling

Once the appropriate techniques have been selected, they must be implemented carefully in a standardized manner. Passive traps create less sampling bias than active collectors do, but even so pitfall traps that are not flush with the ground surface are less effective than those flush with the ground, for example, and are likely to capture fewer small species. The actual location of traps within a habitat may make a considerable difference to the numbers of specimens trapped, depending on vegetation cover, local wind patterns, and so on. Some of these variations (whether inadvertent or not) may actually be advantageous if enough traps are deployed, because they help to characterize trap-by-trap variation for comparison with real site-to-site differences. Again, the most desirable number and pattern of trapping depends on the target groups and habitats selected in accordance with project objectives. Emptying some kinds of traps (*e.g.* emergence traps) requires knowledge and practice so that part of the catch is not missed. Therefore, to reduce avoidable sampling errors, technical staff has to be adequately trained in the placement and servicing of traps before the project begins.

Sorting and Sample Preparation

The sorting and preparation of trap samples of insects is extremely time consuming, taking up to 40 times as long as the sampling itself, even before any identifications are undertaken. Moreover, people differ considerably in the time taken to sort a given sample. Depending on experience, different sorters took from 40 minutes to 5.7 hours for removal of specimens of three major groups from one flight-intercept trap sample, in one instance. As with sampling itself, strict control and standardization are required to ensure the integrity of data and the long-term preservation of specimens. Differences among technicians (Corbet 1966) create additional problems in subsampling or other quantitative work. Constant care and attention is required even during prolonged and relatively mundane activity. For example, laboratory protocols (such as dealing with only one trap sample at a time) must be established to avoid cross-contamination of samples or mislabelling of specimens. Adequate preservative-to-specimen volume must be maintained (by increasing container size if necessary when the trap sample is larger) when field preservative is replaced for more permanent storage. It is particularly important to plan sorting and preparation with identification in mind. Specimens of many groups are much more difficult or impossible to identify if carelessly or incorrectly prepared. Detailed instructions about preparation must therefore be obtained from cooperating systematists as the project is being developed. Of course, groups that require extremely costly preparation (*e.g.* detailed dissection and slide mounting) would need higher relevance to the project objectives to justify the cost of mounting.

Identification

Securing reliable identification to species is the greatest single difficulty in work on insect biodiversity. Except in the few best known groups, expert knowledge is required to ensure that identifications are accurate, and such expertise has to be planned for from the earliest stages, because it is both extremely limited and in great demand for these and many other activities. Despite the increased interest in biodiversity in recent years, the numbers of professional systematists have declined, so that the resources to provide identifications for biodiversity studies simply do not exist. In other words, there is no "black box" into which specimens can be fed for identification. Because a good taxonomist is a scientist, not a technician, specialists are best able to help if a project has a finite objective and is well planned with their participation. Many details of working with relevant systematists can be optimized to favour identification.

Data Management

Studies of biodiversity generate extensive data, requiring efficient means of keeping track of the information and making it available. The necessary protocol at the core of data management keeps basic data associated with each specimen and hence with each species. However, the data (rather than the specimens themselves, despite their taxonomic and voucher value) are used for analyses of biodiversity. It is therefore especially important to consider aspects of data management during the planning stages, so that a system can be put into place before any data are collected. Fortunately, especially because technology has advanced, many specific database systems can be used as long as the data are properly recorded in a logical manner. Provided the computerization has been carefully planned, information in a well organized database then can be extracted using a variety of different formats. A scheme for data management is best developed by considering common or "standard" specimen data, possible additional specific data required for the project at hand, and in a preliminary way how the data will be analysed. Such a preliminary examination helps to avoid inefficient subsequent transcription or splitting of data fields for analysis.

Although standardization of the initial data collected maximizes the availability of information, it is very difficult to ensure because of differences in the capture and subdivision of data fields; subsequent work is hindered by computer incompatibilities such as differences in database structure, vocabularies or hardware configurations. However, progress has been made toward developing standard data fields although some flexibility is needed to respond to unforeseen additional inputs or requirements.

Curation and Disposition of Specimens

The specimens obtained from any study document its findings for future reference, allowing both future checking of data and further use. Data in biology essentially are kept and organized on the basis of the species. Consequently, voucher specimens are required to validate the specific entities obtained during biodiversity studies, because advances in taxonomy (or errors in original identifications) may require re-examination of the material. These voucher specimens also increase progressively in value as knowledge accumulates and they are used for taxonomic or ecological comparison. Such repositories are of considerable long-term importance for evaluating changes or Depending on the objective of a study, vouchers often are best kept associated with the original sites or habitats rather than scattered in a taxonomically arranged collection. However, such a scheme is time consuming and expensive, again reinforcing the need for a long-term plan for the execution and funding of all aspects of a biodiversity study. When most species can be identified (but not as usefully otherwise) the data can be kept associated instead in an electronic format.

Publication and Dissemination of Information

Any properly planned scientific project leads to information that is valid, organized and available. In most cases, this means that results should be peer reviewed and properly published, and not take the form solely of internal reports, unedited lists, and so forth. Information made available on the basis of carefully validated specific identifications is most useful.Information can also be disseminated through computer databases, online information systems, or the Internet. Such dissemination is useful to engender and assist further work and coordination. However, it is very important that the status of any identifications, conclusions or claims that have not been fully evaluated through the focus of expert judgement or peer review and publication be clearly marked as interim.

Conclusions

Five major conclusions emerge from this overview.

1. Any project needs a specific target, choosing the taxonomic and habitat diversity to be sampled through a focus on the project objectives. Indiscriminate or unplanned sampling will not provide answers to relevant questions.
2. Any study must start with a long-term overview of requirements for the whole project, including all stages, from initial planning and wide consultation with a variety of experts, to the disposition of specimens and publication of results.
3. Data of most value for both short-term project and long-term biodiversity goals are based on reliable identification to species. Overcoming the taxonomic impediment for insects therefore is the greatest single challenge for biodiversity studies.
4. Consequently, a key project requirement in addition to funds and their continuity is obtaining systematics expertise. In many cases, such expertise for identification (and indeed for the research required to make identification possible in the groups of most importance to the project objectives) will have to be funded as an explicit and integral component of the project.
5. In turn, the need for taxonomic expertise also requires planning at a level above that of individual projects, involving governmental support, university training, and other infrastructures for systematics.

Insect Biodiversity

In **agricultural ecosystems**, biodiversity is instrumentally important not only for the production of food, but for other ecological services as well, including the recycling of nutrients, regulation of microclimate and local hydrological processes, suppression of undesirable organisms and detoxification of noxious chemicals.

In the United States alone, pollination by bees accounts for over US$9 billion of economic revenue . According to some estimates, over S! of the human diet can be traced directly or indirectly to bee pollination. Losses of key pollinators have been reported in at least one region or country on every continent except Antarctica, which has no pollinators. The Millennium Ecosystem Assessment concluded that with the global decline in the amount of pollinators, there is not a complete loss of fruit or seeds, but a significant decrease in quantity and viability in fruits, and a lower number of seeds.

Food

Over 1000 species of insects are, or have been used as food somewhere in the world. Entomophagy is common in central and southern Africa, Asia, Australia, and Latin America.Termites, crickets, grasshoppers, locusts, beetles, ants, bee brood, and moth larvae are examples of insects that are used as food sources. Insects are high in protein, energy, and a number of minerals and vitamins and can form up to 5-10 per cent of the annual animal protein consumption of certain indigenous groups. Many insects are said to have a nutty flavour, along with a high nutritional content.

Human Culture

Early human civilisations held insects as an important element within their cultures. Most famously, scarab beetles were central religious artifacts within Egyptian culture. **Insect symbolism**, such as the dung beetle portrayed as a potter, is seen in the East as well. The Chinese viewed cicadas as a symbol of birth or immortality, the San of the Kalaharibelieve that the praying mantis represents creation and patience, and the **Greeks** also created beautiful representations of scarab beetles using colourful stones.

The Scope

Over 1 million species of insects have been described, but current estimates of total insect diversity vary from 5-80 million species of insect. Beetles (Coleoptera) make up 40 per cent of described insect species, but some entomologists suggest that flies (Diptera) and Hymenoptera (wasps, bees and ants) could be as diverse or more so. Five orders of insects stand out in their levels of species richness: Hymenoptera, Diptera, Coleoptera, Lepidoptera, and the Hemiptera.

Conservation

There are two approaches to the conservation of insects. Either humans set aside large portions of land using "wilderness preservation" as the motive, or confronting the particular processes that affect the charismatic vertebrates in order to achieve indirect conservation of insects. With biodiversity loss being a global problem, conserving habitat simply for species of insects is of low priority in the current environmental culture.

Single-species conservation is said to preserve many other species indirectly, this preservation by default is referred to as the **umbrella effect.** "Charismatic species", such as butterflies or large, colourful beetles, called flagship species, can expand public awareness and financial contributions for conservation efforts.

Migratory species, such as the well-known monarch butterfly (*Danaus plexippus*), are in need of special conservation methods. One species may require several habitat locations for different periods in their migratory patterns.

Insect conservation has been labeled in the past as a concern only for the affluent. The developing country of Papua New Guinea has a "happily ever after" ending in their attempts to preserve the world's largest butterfly, **Queen Alexandra's Birdwing** (*Ornithoptera alexandrae*). This species is restricted to a very small range of habitat due to specificity in their diet. In the international market of insect collecting, the butterfly can retrieve up to US$2000. In 1978, the government of Papua New Guinea set up the Insect Farming and Trading Agency (IFTA) to regulate the exploitation and conservation of Queen Alexandra's Birdwing and other valuable butterflies.

Assessment of Insect Diversity along an Altitudinal Gradient in Pinderi Forests of Western Himalaya, India

Insect diversity, richness and abundance were evaluated at different altitudes in three forest habitats in the Western Himalayas. The habitats studied were all situated between 2100 and 3500 m and included a site with no disturbances, a site with a moderate level of disturbance and a site with a very high level of disturbance. The species composition and diversity of insects varied at all the three study sites, which demonstrate the effect of altitude and disturbances, as well as the effects of other ecological and climatic parameters on insect populations. The site at lowest altitude, which contained a moderate level of disturbance, supported the highest number of species, whereas the site at the highest altitude, which contained the maximum level of disturbance, supported the lowest number of species. When all of the sites were considered, 122 species of insects belonging to 43 families and 8 orders were recorded. Lepidoptera was the most dominant insect order recorded, with 46 species being observed. This was followed by Hymenoptera (20), Coleoptera (18), Orthoptera (12), Hemiptera (10), Odonata (9), Diptera (5) and Dermeptera (2). The most abundant species were *Vanessa cashmirensis* Fru. (Lepidoptera: Nymphalidae), *Pieris canidia indica* Sparr. (Lepidoptera: Pieridae), *Apis laboriosa* Smith (Hymenoptera: Apidae), *Anomala dimidiata* Hope (Coleoptera: Scarabidae), *Chorthipus* sp. (Orthoptera: Acrididae), *Crocothemis servilia servilia* (Drury) (Odonata: Libellulidae) and *Syrphus fulvifacies* Brunetti (Diptera: Syrphidae). The site at the lowest altitude and the sites with the longest rainy seasons had the highest Shannon–Wiener Diversity.

Diversity Stability Debate

There exists little doubt that the Earth's biodiversity is declining. The Nature Conservancy, for example, has documented that one-third of the plant and animal species in the United States are now at risk of extinction. The problem is a monumental one, and forces us to consider in depth how we expect ecosystems, which ultimately are our life-support systems, to respond to reductions in diversity. This issue – commonly referred to as the diversity–stability debate – is the subject of this review, which synthesizes historical ideas with recent advances. Both theory and empirical evidence agree that we should expect declines in diversity to accelerate the simplification of ecological communities.

We now realize that the world's flora and fauna are disappearing at rates greater than the mass extinction events whose collapses punctuate the fossil record. It is also true that species invasions have been elevated to unprecedented rates accompanying the increased globalization of our world. These high rates of extinction and invasion put ecosystems under enormous stress, making it critical that we understand how the loss, or addition, of a species influences the stability and function of the ecosystems we rely on. We are, in a very real sense, deconstructing the Earth under the implicit assumption that ecosystems have evolved the ability to withstand such assault without collapse.

Several advances in the diversity–stability debate form a conceptual thread that suggests that diversity can be expected, on average, to give rise to ecosystem stability. This does not infer that diversity is the driver of this relationship. Instead, diversity can be regarded as the passive recipient of important ecological mechanisms that are inherent in ecosystems. One promising mechanism that has been proposed recently is that weakly interacting species stabilize community dynamics by dampening strong, potentially destabilizing consumer–resource interactions6. Empirical descriptions of the distribution of interaction strengths in real communities are consistent with this theory. If this is true then, all else being equal, decreasing biodiversity will be accompanied by increases in average interaction strengths within ecosystems, and a concomitant decrease in ecosystem stability.

Historical Perspectives of the Diversity–Stability Debate

The relationship between diversity and stability has fascinated ecologists. Before the 1970s, ecologists believed that more diverse communities enhanced ecosystem stability. A strong proponent of this view was Charles Elton, who argued that "simple communities were more easily upset than that of richer ones; that is, more subject to destructive oscillations in populations, and more vulnerable to invasions". In fact, both Odum and Elton8 arrived at similar conclusions based on the repeated observation that greatly simplified terrestrial communities are characterized by more violent fluctuations in population density than diverse terrestrial communities. For example, invasions most frequently occur on cultivated land where human influence had produced greatly simplified ecological communities, and outbreaks of phytophagous insects occur readily in boreal forests but are unheard of in diverse tropical forests. These observations led Elton to believe that complex communities, constructed from many predators and parasites, prevented populations from undergoing explosive growth. His ideas were closely akin to MacArthur, who reasoned that multiplicity in the number of prey and predator species associated with a population freed that population from dramatic changes in abundance when one of the prey or predator species declined in density.

These early intuitive ideas were challenged by the work of Robert May in 1973. May turned to mathematics to rigorously explore the diversity–stability relationship. By using linear stability analysis on models constructed from a statistical universe (that is, randomly constructed communities with randomly assigned interaction strengths), May found that diversity tends to destabilize community dynamics. Other ecologists, using similar approaches, found results that were consistent with this hypothesis. The results were puzzling, as real ecosystems were undoubtedly complex and diverse. The results also seemed to counter the ideas of Elton, Odum and MacArthur. Yodzis heightened this paradox by showing that models structured from compiled food-web relationships, with plausible interaction strengths, were generally more stable than randomly constructed food webs. Although the early food-web data that Yodzis structured his models around were incomplete, these data reflected real feeding relationships. Yodzis' result indicated that interaction strength was probably crucial to stability; but the exact reason for this remained elusive. If diversity and stability were positively correlated, as early empirical evidence had indicated, then more had to be happening than simply increasing the number of species and the number of pathways. Something fundamental was missing from the early arguments.

In the remainder of this paper I review recent investigations of the diversity–stability debate. I first discuss a change in perspective that is beginning to allow us to unravel this long-standing problem and then review two different lines of investigation. One approach has searched for a general diversity–stability relationship, and a second, more mechanistic approach has sought a relationship between food-web structure and stability.

Changing Perspectives

Much of ecological theory is based on the underlying assumption of equilibrium population dynamics. Although this assumption is aesthetically pleasing, in that it suggests the balance of nature is infinitely precise, an alternative and viable ecological perspective exists. As real populations are variable, it is possible that the persistence of complex communities depends to some degree on population fluxes (that is, the fairly regular waxing and waning of a population's density). Such background population variability, whether driven by biotic or abiotic processes, can provide species with the opportunity to respond differentially to their environment. In turn, these differential species responses weaken the destructive potential of competitive exclusion.

Because such variability can significantly change our understanding of ecological interactions, ecologists have begun to relax equilibrium-based measures of stability. A recent theoretical analysis has shown that population fluctuations, driven by competition, can actually promote the persistence of large numbers of competing phytoplankton communities on a minimal number of limiting resources (but greater than two resources). Coexistence was found to rely on the fluctuation in population densities, while community-level densities (the summation of the competing plankton densities) varied little. We will see that a similar relationship appears in diversity–stability experiments. Here, too, the evidence points to variable population densities that sums to produce a relatively constant biomass at the community level.

Definitions of Stability

Definitions of stability in ecology can be classified generally into two categories – stability definitions that are based on a system's dynamic stability and stability definitions that are based on a system's ability to defy change. Despite the breadth of definitions, ecological theory has tended traditionally to rely on the assumption that a system is stable if, and only if, it is governed by stable equilibrium dynamics (that is, equilibrium stability and equilibrium resilience). As discussed in the previous section, these are strong assumptions with no *a priori* justification. In fact, the variable nature of population dynamics found both in field and in laboratory experiments has led experimentalists to use measures of variability as indices of a system's stability. This discontinuity between stability experiments and equilibrium-based theory has made it difficult to unite theory and experiment in the diversity–stability debate.

More general definitions of stability exist. In, general stability is defined such that stability increases as population densities move further away from extremely low or high densities. This is a broad definition, including equilibrium and non-equilibrium dynamics as well as subsuming the definition of permanence (a population is considered permanent if the lower limit to its density is greater than zero). Because the definition of general stability implies decreased variability (owing to greater limits on density), it is closely related to field measurements of stability, which tend to rely on variability in population or community densities as a measure of stability. One can also extend equilibrium resilience to a less biologically restrictive form by defining resilience as the return time after a perturbation to an equilibrium or a non-equilibrium attractor. In a nonlinear system there is no reason to believe that an equilibrium that attracts weakly in a local setting (near the equilibrium) also attracts weakly far away from the equilibrium, where the issue of a species' permanence is resolved. For the remainder of the paper, unless stated otherwise, the definitions of general stability and variability will be used to consider empirical and theoretical results on the relationship between diversity and stability under a common framework.

The Search for a General Diversity–Stability Relationship

In 1982, David Tilman began a long-term study to delineate experimentally the relationship between diversity and stability in plant communities. The undertaking involved four grassland fields at Cedar Creek Natural History Area, Minnesota, divided into over 200 experimental plots, and gathered information on species richness, community biomass and population biomass through time. The results of this and other extensive studies converge on the finding that diversity within an ecosystem tends to be correlated positively with plant community stability (that is, decreased coefficient of variability in community biomass). At the same time, diversity seems to show little influence on population variability. The basic arguments for a positive relationship between diversity and stability for primary producers at the community level have been classified into two, not mutually exclusive, hypotheses called the averaging effect and the negative covariance effect. In essence, these hypotheses argue that diversity (species richness) increases stability at the community level because diverse plant communities respond differentially to variable background processes. The differential responses of populations sum, through time, to give stable community dynamics.

If diversity and stability are positively correlated, then both the averaging and negative covariance effect predicts that population variance has to scale as a function of mean population densities in a precise way.Tilman has used these predictions to show that his field experiments are consistent with the interpretation that increasing diversity increases community stability. Although this is a clever combination of theory and experiment, it cannot be used to infer that diversity is responsible directly for stability. As a counter example, no correlation was found between diversity and stability at the cross-ecosystem scale. Other experiments have found that the positive diversity–stability correlation is not a pure species effect (that is, a diversity effect), and have indicated that ecosystem function and stability are more directly related to functional diversity (for example, graminoids or grasses, nitrogen-fixing legumes and other herbs).

In a similar manner, plant community stability and productivity in European grasslands were shown to be tightly coupled to the functional diversity of mutualistic arbuscular mycorrhizal fungi (AMF). In this system, large fluctuations in plant biomass were associated with low diversities of AMF, whereas more constant biomass and greater productivity accompanied high AMF diversities. This study highlights that higher-level interaction, which is inherent in food webs (for example, microbial interactions, herbivory and predation), are of great importance in understanding the relationship between the diversity and stability of whole ecological communities. The complexity of whole ecological communities – the basis from which Odum, Elton and MacArthur formed their diversity–stability hypotheses – cannot manifest itself in experiments that focus on single trophic levels.

Field tests at the scale of the food web are few in number. But one thorough examination tested seven different stability–diversity criteria in the grazing ecosystem of the Serengeti under naturally variable conditions (that is, strong seasonal changes). Of these seven stability measures, five were positively related to diversity whereas two were unrelated to diversity. The study found that greater diversity reduced the relative magnitudes of fluctuations in productivity induced by seasonal change. Although a relationship between stability and diversity exists within the Serengeti, the evidence again points to the importance of functional species in understanding this relationship. For example, the grazing-tolerant plant species have a disproportionately large role in the Serengeti community dynamics by preventing herbivores from dramatically reducing plant biomass.

The paucity of field tests at the scale of the food web reflects the fact that such experiments require an enormous undertaking. As an alternative, ecologists have approached this problem by investigating how diversity influences stability and function within a multitrophic setting in controlled microcosm experiments (often referred to as bottle experiments as they are attempts to create realistic ecological communities within a controlled setting). The main advantage of microcosms is that the experiments can easily be manipulated and replicated. Nonetheless, the issue of how scale influences outcome looms over microcosm experiments – can we extrapolates results to the whole ecosystem? Ambitious experimental set ups such as that currently underway in the Ecotron are attempting to bridge the gap between the complexity of real field communities and the simplicity of laboratory or greenhouse experiments.

Conclusion

Taken together, recent advances indicate that diversity can be expected, on average, to give rise to ecosystem stability. The evidence also indicates that diversity is not the driver of this relationship; rather, ecosystem stability depends on the ability for communities to contain species, or functional groups that are capable of differential response. All of these views are consistent with the ideas put forth by the influential figures of **Odum, Elton, MacArthur and May**. May's result reflects the fact that random distributions (that is, a null universe where dynamics are influenced by diversity alone), on average, do not create the necessary tension between community members that forces differential response and community stability. In a randomly constructed community, for example, strong interactions are not necessarily coupled to weak interactions that mute their destabilizing potential. In fact, one can expect that random communities will generally not create such couplings, and so tend to produce diverse communities with complex, wildly oscillatory dynamics. Furthermore, Odum, Elton and MacArthur recognized that real food webs contain a complex array of energetic pathways that can act as buffers against dramatic population explosions. Specifically, MacArthur's hypothesis – that greater connectance drives community and ecosystem stability – seems a strong possibility provided most pathways are constructed from weak interactions that mute the potentially destabilizing roles of a few strong consumer–resource interactions.

The current empirical evidence indicates that communities may be dominated by such weak trophic interactions. If this is true, then it is also true that the removal, or addition, of any species (weak or strong) can lead to pronounced changes in community composition and structure. It follows that decreasing biodiversity will tend to increase the overall mean interaction strength, on average, and thus increase the probability that ecosystems undergo destabilizing dynamics and collapses. Just how much ecosystem deterioration is sufficient to precipitate a collapse is difficult to assess, but current experiments and theory agree that drastic community changes can accompany the removal or addition of even a single species. Furthermore, if Elton's observation is correct – that simplified communities are more vulnerable to invasion – than we should also expect an increase in frequency of successful invaders as well as an increase in their impact as our ecosystems become simplified. The lessons for conservation are obvious: (1) if we wish to preserve an ecosystem and its component species then we are best to proceed as if each species is sacred; and (2) species removals (that is, extinction) or species additions (that is, invasions) can, and eventually will, invoke major shifts in community structure and dynamics.

It is important to point out that the mechanistic ideas outlined here omit higher-scale ecosystem influences that are likely to be linked intricately to ecosystem stability and function. Some promising work is now beginning to show us how we can link models of nutrient and energy flow as well as uncover the potential influence of diversity and stability on large-scale biogeochemical processes. Investigations of this sort will be necessary to bridge important stabilizing processes that act across ecological scales.

Relevance to Pest Management

Integrated pest management (IPM), also known as Integrated Pest Control (IPC) is a broad-based approach that integrates practices for economic control of pests. IPM aims to suppress pest populations below the economic injury level (EIL). The UN's **Food and Agriculture Organisation** defines IPM as "the careful consideration of all available pest control techniques and subsequent integration of appropriate measures that discourage the development of pest populations and keep pesticides and other interventions to levels that are economically justified and reduce or minimize risks to human health and the environment. IPM emphasizes the growth of a healthy crop with the least possible disruption to agro-ecosystems and encourages natural pest control mechanisms." **Entomologists** and **ecologists** have urged the adoption of IPM pest control since the 1970s. IPM allows for safer pest control. This includes managing insects, plant pathogens and weeds.

Globalization and increased mobility open allow increasing numbers of invasive species to cross national borders. IPM poses the least risks while maximizing benefits and reducing costs.

For their leadership in developing and spreading IPM worldwide, Perry Adkisson and Ray F. Smith received the 1997 **World Food Prize**.

History

Shortly after World War II, when synthetic insecticides became widely available, entomologists in California developed the concept of "supervised insect control". Around the same time, entomologists in the US Cotton Belt were advocating a similar approach. Under this scheme, insect control was "supervised" by qualified entomologists and insecticide applications were based on conclusions reached from periodic monitoring of pest and natural-enemy populations. This was viewed as an alternative to calendar-based programs. Supervised control was based on knowledge of the ecology and analysis of projected trends in pest and natural-enemy populations.

Supervised control formed much of the conceptual basis for the "integrated control" that University of California entomologists articulated in the 1950s. Integrated control sought to identify the best mix of chemical and biological controls for a given insect pest. Chemical insecticides were to be used in the manner least disruptive to biological control. The term "integrated" was thus synonymous with "compatible." Chemical controls were to be applied only after regular monitoring indicated that a pest population had reached a level (the economic threshold) that required treatment to prevent the population from reaching a level (the economic injury level) at which economic losses would exceed the cost of the control measures.

IPM extended the concept of integrated control to all classes of pests and was expanded to include all tactics. Controls such as pesticides were to be applied as in integrated control, but these now had to be compatible with tactics for all classes of pests. Other tactics, such as host-plant resistance and cultural manipulations, became part of the IPM framework. IPM combined entomologists, **plant pathologists, nematologists** and weed scientists.

In the United States, IPM was formulated into national policy in February 1972 when President Richard Nixon directed federal agencies to take steps to advance the application of IPM in all relevant sectors. In 1979, President Jimmy Carter established an interagency IPM Coordinating Committee to ensure development and implementation of IPM practices.

Application

IPM is used in **agriculture, horticulture,** human habitations, **preventive conservation** and general pest control, including structural pest management, turf pest management and ornamental pest management.

Principles

An American IPM system is designed around six basic components:

- **Acceptable pest levels**–The emphasis is on *control,* not *eradication.* IPM holds that wiping out an entire pest population is often impossible, and the attempt can be expensive and unsafe. IPM programmes first work to establish acceptable pest levels, called action thresholds, and apply controls if those thresholds are crossed. These thresholds are pest and site specific, meaning that it may be acceptable at one site to have a weed such as white clover, but not at another site. Allowing a pest population to survive at a reasonable threshold reduces selection pressure. This lowers the rate at which a pest develops resistance to a control, because if almost all pests are killed then those that have resistance will provide the genetic basis of the future population. Retaining significant number unresistant specimens dilutes the prevalence of any resistant genes that appear. Similarly, the repeated use of a single class of controls will create pest populations that are more resistant to that class, whereas alternating among classes helps prevent this.
- Preventive cultural practices–Selecting varieties best for local growing conditions and maintaining healthy crops is the first line of defense. Plant quarantine and 'cultural techniques' such as crop sanitation are next, *e.g.*, removal of diseased plants, and cleaning pruning shears to prevent spread of infections. Beneficial fungi and bacteria are added to the potting media of horticultural crops vulnerable to root diseases, greatly reducing the need for fungicides.
- **Monitoring**–Regular observation is critically important. Observation is broken into inspection and identification. Visual inspection, insect and spore traps, and other methods are used to monitor pest levels. Record-keeping is essential, as is a thorough knowledge target pest behavior and reproductive cycles. Since insects are cold-blooded, their physical development is dependent on area temperatures. Many insects have had their development cycles modeled in terms of **degree-days**. The degree days of an environment determines the optimal time for a specific insect outbreak. Plant pathogens follow similar patterns of response to weather and season.

- ✰ **Mechanical controls**–Should a pest reach an unacceptable level, mechanical methods are the first options. They include simple hand-picking, barriers, traps, vacuuming andtillage to disrupt breeding.
- ✰ **Biological controls**–Natural biological processes and materials can provide control, with acceptable environmental impact, and often at lower cost. The main approach is to promote beneficial insects that eat or parasitize target pests. Biological insecticides, derived from naturally occurring microorganisms (*e.g.–Bt*, entomopathogenic fungi andentomopathogenic nematodes), also fall in this category. Further 'biology-based' or 'ecological' techniques are under evaluation.
- ✰ **Responsible use**–Synthetic pesticides are used as required and often only at specific times in a pest's life cycle. Many newer pesticides are derived from plants or naturally occurring substances (*e.g.*–nicotine, pyrethrum and insect juvenile hormone analogues), but the toxophore or active component may be altered to provide increased biological activity or stability. Applications of pesticides must reach their intended targets. Matching the application technique to the crop, the pest, and the pesticide is critical. The use of low-volume spray equipment reduces overall pesticide use and labor cost.

An IPM regime can be simple or sophisticated. Historically, the main focus of IPM programmes was on agricultural insect pests. Although originally developed for agricultural pest management, IPM programmes are now developed to encompass diseases, weeds and other pests that interfere with management objectives for sites such as residential and commercial structures, lawn and turf areas, and home and **community gardens**.

Process

IPM is the selection and use of pest control actions that will ensure favourable economic, ecological and social consequences and is applicable to most agricultural, public health and amenity pest management situations. The IPM process starts with monitoring, which includes inspection and identification, followed by the establishment of economic injury levels. The economic injury levels set the economic threshold level. That is the point when pest damage (and the benefits of treating the pest) exceed the cost of treatment. This can also be an action threshold level for determining an unacceptable level that is not tied to economic injury. Action thresholds are more common in structural pest management and economic injury levels in classic agricultural pest management. An example of an action threshold is one fly in a hospital operating room is not acceptable, but one fly in a pet kennel would be acceptable. Once a threshold has been crossed by the pest population action steps need to be taken to reduce and control the pest. Integrated pest management employ a variety of actions including cultural controls, including physical barriers, biological controls, including adding and conserving natural predators and enemies to the pest, and finally chemical controls or pesticides. Reliance on knowledge, experience, observation and integration of multiple techniques makes IPM appropriate for organic farming (excluding synthetic pesticides). These may or may not include materials listed on the Organic Materials Review Institute (OMRI) Although the pesticides and particularly insecticides used in organic farming and organic gardening are generally safer than synthetic pesticides, they are not always more safe or environmentally friendly than synthetic pesticides and can cause harm. For conventional farms IPM can reduce human and environmental exposure to hazardous chemicals, and potentially lower overall costs.

Risk assessment usually includes four issues: 1) characterization of biological control agents, 2) health risks, 3) environmental risks and 4) efficacy.

Mistaken identification of a pest may result in ineffective actions. *E.g.*, plant damage due to over-watering could be mistaken for fungal infection, since many fungal and viral infections arise under moist conditions.

Monitoring begins immediately, before the pest's activity becomes significant. Monitoring of agricultural pests includes tracking soil/planting media fertility and water quality. Overall plant health and resistance to pests is greatly influenced by pH, alkalinity, of dissolved mineral and Oxygen Reduction Potential. Many diseases are waterborne, spread directly by irrigation water and indirectly by splashing.

Once the pest is known, knowledge of its lifecycle provides the optimal intervention points. For example weeds reproducing from last year's seed can be prevented with mulches and pre-emergent herbicide.

Pest-tolerant crops such as soybeans may not warrant interventions unless the pests are numerous or rapidly increasing. Intervention is warranted if the expected cost of damage by the pest is more than the cost of control. Health hazards may require intervention that is not warranted by economic considerations.

Specific sites may also have varying requirements. *E.g.*, white clover may be acceptable on the sides of a tee box on a golf course, but unacceptable in the fairway where it could confuse the field of play.

Possible interventions include mechanical/physical, cultural, biological and chemical. Mechanical/physical controls include picking pests off plants, or using netting or other material to exclude pests such as birds from grapes or rodents from structures. Cultural controls include keeping an area free of conducive conditions by removing waste or diseased plants, flooding, sanding, and the use of disease-resistant crop varieties. Biological controls are numerous. They include: conservation of natural predators or augmentation of natural predators, **sterile insect technique** (SIT).

Augmentation, inoculative release and inundative release are different methods of biological control that affect the target pest in different ways. Augmentative control includes the periodic introduction of predators. With inundative release, predators are collected, mass-reared and periodically released in large numbers into the pest area. This is used for an immediate reduction in host populations, generally for annual crops, but is not suitable for long run use. With inoculative release a limited number of beneficial organisms are introduced at the start of the growing season. This strategy offers long term control as the organism's progeny affect pest populations throughout the season and is common in orchards. With seasonal inoculative release the beneficials are collected, mass-reared and released seasonally to maintain the beneficial population. This is commonly used in greenhouses. In America and other western countries, inundative releases are predominant, while Asia and the Eastern Europe more commonly use inoculation and occasional introductions.

The **Sterile insect technique** (SIT) is an Area-Wide IPM program that introduces sterile male pests into the pest population to trick females into (unsuccessful) breeding encounters, providing a form of birth control and reducing reproduction rates. The biological controls mentioned above only appropriate in extreme cases, because in the introduction of new species or supplementation of naturally occurring species can have detrimental ecosystem effects. Biological controls can be used to stop invasive species or pests, but they can become an introduction path for new pests.

Chemical controls include horticultural oils or the application of insecticides and herbicides. A **Green Pest Management** IPM program uses pesticides derived from plants, such as botanicals, or other naturally occurring materials.

Pesticides can be classified by their modes of action. Rotating among materials with different modes of action minimizes pest resistance.

Evaluation is the process of assessing whether the intervention was effective, whether it produced unacceptable side effects, whether to continue, revise or abandon the program.

Southeast Asia

The Green Revolution of the 1960s and '70s introduced sturdier plants that could support the heavier grain loads resulting from intensive fertilizer use. Pesticide imports by 11Southeast Asian countries grew nearly sevenfold in value between 1990 and 2010, according to FAO statistics, with disastrous results. Rice farmers become accustomed to spraying soon after planting, triggered by signs of the leaf folder moth, which appears early in the growing season. It causes only superficial damage and doesn't reduce yields. In 1986, Indonesia banned 57 pesticides and completely stopped subsidizing their use. Progress was reversed in the 2000s, when growing production capacity, particularly in China, reduced prices. Rice production in Asia more than doubled. But it left farmers believing more is better–whether it's seed, fertilizer, or pesticides.

The **brown planthopper** (Nilaparvata lugens), the farmers' main target, has become increasingly resistant. Since 2008, outbreaks have devastated rice harvests throughout Asia, but not in the Mekong Delta. Reduced spraying allowed natural predators to neutralize planthoppers in Vietnam. In 2010 and 2011, massive planthopper outbreaks hit 400,000 hectares of Thai rice fields, causing losses of about $64 million. The Thai government is now pushing the "no spray in the first 40 days" approach.

By contrast early spraying kills frogs, spiders, wasps and dragonflies that prey on the later-arriving and dangerous planthopper and produced resistant strains. Planthoppers now require pesticide doses 500 times greater than originally. Overuse indiscriminately kills beneficial insects and decimates bird and amphibian populations. Pesticides are suspected of harming human health and became a common means for rural Asians to commit suicide.

In 2001, scientists challenged 950 Vietnamese farmers to try IPM. In one plot, each farmer grew rice using their usual amounts of seed and fertilizer, applying pesticide as they chose. In a nearby plot, less seed and fertilizer were used and no pesticides were applied for 40 days after planting. Yield from the experimental plots was as good or better and costs were lower, generating 8 per cent to 10 per cent more net income. The experiment led to the "three reductions, three gains" campaign, claiming that cutting the use of seed, fertilizer and pesticide would boost yield, quality and income. Posters, leaflets, TV commercials and a 2004 radio soap opera that featured a rice farmer who gradually accepted the changes. It didn't hurt that a 2006 planthopper outbreak hit farmers using insecticides harder than those who didn't. Mekong Delta farmers cut insecticide spraying from five times per crop cycle to zero to one.

The Plant Protection Center and the **International Rice Research Institute** (IRRI) have been encouraging farmers to grow flowers, okra and beans on rice paddy banks, instead of stripping vegetation, as was typical. The plants attract bees and a tiny wasp that eats planthopper eggs, while the vegetables diversify farm incomes.

Agriculture companies offer bundles of pesticides with seeds and fertilizer, with incentives for volume purchases. A proposed law in Vietnam requires licensing pesticide dealers and government approval of advertisements to prevent exaggerated claims. Insecticides that target other pests, such as *Scirpophaga incertulas* (stem borer), the larvae of moth species that feed on rice plants allegedly yield gains of 21 per cent with proper use.

Relevance of Traditional Integrated Pest Management (IPM) Strategies for Commercial Corn Producers in a Transgenic Agroecosystem: A Bygone Era?

The use of transgenic *Bt* maize hybrids continues to increase significantly across the Corn Belt of the United States. In 2009, 59 per cent of all maize planted in Illinois was characterized as a "stacked" gene variety. This is a 40 per cent increase since 2006. Stacked hybrids typically express one Cry protein for corn rootworm control and one Cry protein for control of several lepidopteran pests; they also feature herbicide tolerance (to either glyphosate or glufosinate). Slightly more than 50 years has passed since Vernon Stern and his University of California entomology colleagues published (1959) their seminal paper on the integrated control concept, laying the foundation for modern pest management (IPM) programs. To assess the relevance of traditional IPM concepts within a transgenic agroecosystem, commercial maize producers were surveyed at a series of meetings in 2009 and 2010 regarding their perceptions on their use of *Bt* hybrids and resistance management. Special attention was devoted to two insect pests of corn, the European corn borer and the western corn rootworm. A high percentage of producers who participated in these meetings planted *Bt* hybrids in 2008 and 2009, 97 and 96.7 per cent , respectively. Refuge compliance in 2008 and 2009, as mandated by the U.S. Environmental Protection

Agency (EPA), was 82 and 75.7 per cent , respectively, for those producers surveyed. A large majority of producers (79 and 73.3 per cent in 2009 and 2010, respectively) revealed that they would, or had, used a *Bt* hybrid for corn rootworm (*Diabrotica virgifera virgifera* LeConte) or European corn borer (*Ostrinia nubilalis* Hübner) control even when anticipated densities were low. Currently, the EPA is evaluating the long-term use of seed blends (*Bt* and non-*Bt*) as a resistance management strategy. In 2010, a large percentage of producers, 80.4 per cent , indicated they would be willing to use this approach. The current lack of integration of management tactics for insect pests of maize in the U.S. Corn Belt, due primarily to the escalating use of transgenic *Bt* hybrids, may eventually result in resistance evolution and/or other unforeseen consequences.

Spatial Patterns of Insect Pests in Agricultural Systems

The forecast of insect populationtrends in the temporal direction and the economic impact of population loads upon agricultural production, was one of the central aspects of IPM during the second one half of the previous century. Phenological models were constructed for a vast variety of insects and crops and used to make decisions regarding the application of control measures. Most of these models, however did not take into consideration the spatial and geographic heterogeneity of the environment, a situation which may have resulted in a reduced ability of models to forecast population trends and a lower adoption rate of system approach for the pest management.

The space (e. g. a continuous extension) is the place where living organism interact among themselves and with the physical and chemical environment. In general the dispersion patterns of biological and nonliving elements included within a space tend to be heterogenous at most measure scale. (Levin 1992). That is, the values of measured variable of any element in a continuous extension will differ even at short distances. This heterogeneity in space, and the understanding of the mechanism underlying this phenomenon, is cardinal to our interpretation of the environment and the understanding of the ecology of living organism.

As elements within agricultural systems, insect pests are expected to disperse within the space of the agroecosystem in a heterogenous fashion. Therefore the understanding of the spatial dimension of the agricultural system and the spatial dispersion patterns of insect pests is expected to provide us with an essential tool in IPM.

Chapter 16

Pest Management as Applied Ecology

Incorporating pest controls at many different stages and limiting pests' abilities in many small ways are the foundation of ecological pest management. Production systems that use ecological principles to imitate nature, along with multiple tactics and the right information, can:

- ☆ Strengthen individual impacts of strategies when used together,
- ☆ Reduce the risk of crop failure by distributing the burden of crop protection across many tactics,
- ☆ Minimize environmental disruptions and threats to human health,
- ☆ Slow the rate at which pests adapt or evolve resistance to a given management tactic because that tactic is used less frequently, and
- ☆ Reduce operating costs and improve profitability by minimizing the need for purchased inputs.

Cotton research headed by Joe Lewis at ARS in Georgia has shown that, like Steve Groff's vegetable system, combining minimum tillage with cover crops and cover crop mulch creates enough biological diversity to stymie pests. Comparing tilled fields to fields planted using conservation tillage following a winter cover crop like vetch, winter grains or clover, researchers found that beneficial insect populations increased. In fact, overall seasonal densities of certain types of carabid beetles and spiders in the 'conservation' fields were a full 14 times higher than in the conventional fields.

Input costs were nearly identical, but average yields in the conservation fields were about 100 pounds higher than conventional yields. Moreover, net returns were $60 per acre higher in the conservation plots.

Improve Management of the Disturbances Created by Agriculture

Agricultural disturbances such as tillage, harvest, and fertilizer and pesticide application all can provoke pest problems, but you can avoid stimulating pests at the wrong time. For example, till fields before final seedbed preparation to stimulate weed germination, then cultivate before planting to lower the density of weeds infesting a crop.

Leaving some undisturbed areas on a farm can help maintain the balance between beneficial and pest organisms. Many predators and parasites that attack crop pests thrive in the less-disturbed areas provided by hedgerows, weedy borders, woodlots and riparian buffers on the farm; in grassed alleyways in orchards and grassed waterways in field crops; and even in the small areas left between crop rows by zone tillage. Small sites allow natural enemies to persist and migrate into crop fields to keep pest populations in check.

In a research project in the Southeast, ground beetles, field crickets, ants and field mice were important weed seed predators within a low-input, no-till cropping system in which soybeans were grown in a surface mulch of wheat straw. Over five weeks in the fall, the weed seed predators removed more than double the number of seeds from the no-till system compared to an adjacent conventional tillage system.

Include Perennial Plants in and Near Fields

Perennial plants - such as fruit trees, grassed waterways, trees growing along stream banks, or forage grasses and legumes harvested for hay - offer many advantages:

Their roots are more extensive and longer lasting than those of annual crops,

Much more than annual crops, they support communities of diverse soil organisms that are more similar to those in soils of natural ecosystems,

They enhance water infiltration and reduce soil compaction, thus extending rooting depth,

they serve as important habitat for beneficial insects, providing both food and shelter, and

they help preserve soil and water quality by maintaining living plant cover above ground and active roots in the soil.

Increase Diversity

Diversity, both in the crops you grow and how you manage them, can reduce pest problems, decrease the risks of market and weather fluctuations, and eliminate labor bottlenecks. Enrich diversity: across the landscape (within fields, on the farm as a whole and throughout a local watershed), throughout the season (different crops on the same farm at different stages of growth and managed in different ways), and from year to year (rotations of three or more crops).

Ideally, agricultural landscapes will look like patchwork quilts: dissimilar types of crops growing at various stages and under diverse management practices. Within this confusing patchwork, pests will encounter a broader range of stresses and will have trouble locating their hosts in both space and time. Their resistance to control measures also will be hampered.

As plant diversity intensifies above ground, diversity builds in the soil. Through a system of checks and balances, a medley of soil organisms helps maintain low populations of many pests. Good soil tilth and generous quantities of organic matter also can stimulate this very useful diversity in pest-fighting soil organisms.

Researcher Matt Liebman reviewed cropping system studies to get at how plant diversity deters weeds. His summary of various studies that grew 27 test crops in rotation compared to monoculture systems found that:

- Weed plant density in rotation was less than in monoculture in 19 out of 25 cases,
- Weed seed density in crop rotation was lower in 9 out of 12 cases, and
- Yields of test crops were higher in rotation than monoculture in 9 out of 12 cases.

'These results suggest that crop rotation can be an important component of strategies to reduce weed density and maintain or increase crop yield,' Liebman says.

In Oregon's Willamette Valley, Larry Thompson's 100-acre fruit and vegetable farm blossoms with natural insectaries. 'To keep an equilibrium of beneficials and pests and to survive without using insecticides, we have as much blooming around the farm as we can,' he says.

Thompson uses cover crops to recruit ladybugs, lacewings and praying mantises in his battle against aphids. Overseeded cereal rye is already growing under his lettuce leaves before he harvests in late summer and fall. 'It creates a nice habitat for overwintering beneficials and you don't have to start over from ground zero in the spring,' he says.

Between his raspberry rows, Thompson lets his dandelions flower into a food source for nectar- and pollen-seeking insects before mowing them down. Forced out of the dandelions that nurtured them in early spring, the beneficials pursue a succession of bloom. They move first into his raspberries, then his Marion berries and boysenberries.

Later in the year, Thompson doesn't mow his broccoli stubble. Instead, he lets the side shoots bloom, creating a long-term nectar source into early winter. 'The bees really go for that,' he says.

Reducing pest pressure

Reducing pest problems relies on many 'little hammers,' each contributing to one or more of the following general strategies:

- ☆ Producing healthy crops
- ☆ Increasing stress on pests
- ☆ Enhancing beneficials

You can redesign the farm to become a more complex agricultural ecosystem. Maximize the farm's favorable ecological processes, such as nitrogen fixation, nutrient mineralization from organic matter and beneficial insect populations. Minimize undesirable processes, such as nutrient loss, disease development and feeding damage by crop pests.

'We're not trying to turn farms into completely natural systems,' says Teasdale. 'In a natural system, no one species becomes dominant. In an agroecosystem, the crop is going to dominate. But within that much simpler, very managed system, we can apply many principles from natural ecosystems to make it easier to control pests.'

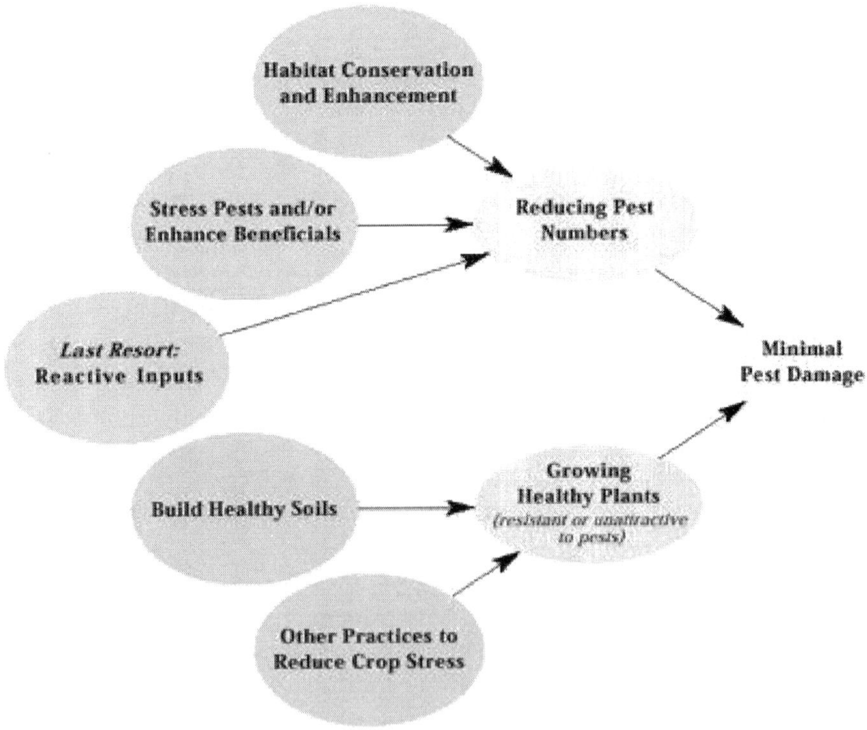

Produce Healthy Crops

Vigorous crops compete better with weeds and tolerate more insect damage and disease. Growing crop varieties that are resistant to particular pests, such as a fungal disease, usually results in more vigorous crops that are better able to resist other pests. Reducing environmental stresses through better soil and crop management helps plants better compete with or resist pests.

Build and Maintain Soil Health

The link between healthy soils and healthy plants remains fundamental to ecologically based pest management. The ability of a plant to resist or tolerate pests is grounded in favorable physical, chemical and biological properties of soil. Adequate moisture, good soil tilth, appropriate pH, the right amounts and balance of nutrients, and a diverse and active community of soil organisms all contribute to plant health.

Conserving and building soil organic matter encourages soil fertility and promotes more complex food webs among soil organisms. Healthy plants depend on healthy root systems. 'Creating aerobic soil conditions increases the health of plant roots,' says John Luna, extension specialist in integrated farming systems at Oregon State University. 'By maintaining good drainage, good tilth and good aerobic condition, you're able to promote a whole array of beneficial microorganisms and to discourage the pestiferous ones.'

Beneficial bacteria and fungi that colonize root surfaces can prevent infection by such disease-causing organisms as *Pythium* and *Rhizoctonia*, especially in biologically diverse systems with more complex food webs. Beneficial soil fungi, nematodes and insects also can be more effective in complex than in simple soil systems.

Scientists are finding that contact with pest invaders can actually mobilize resistance mechanisms in plants. For example, a leaf infection by a plant pathogen or a bite by an insect can prompt resistance to future attacks by these or very different pests. This 'systemic acquired resistance' occurs throughout the plant, even in tissue far away from the initial site. Farming practices can enhance it. Amending soil with compost, for example, has produced systemic resistance within cucumber to anthracnose. Similarly, inoculating transplants with beneficial mycorrhizal fungi has protected roots from root rot fungi such as *Cylindrocarpon* and *Pythium*.

Composting imported organic-waste residues before applying them to soils may help fight crop diseases. Good composts are costly to buy and slow to produce, but they can pay their own way - especially on farms that produce high-value vegetables and small-berry fruits. At Ohio State University, plant pathologist Harry Hoitink and his co-workers have found that compost may suppress root and foliar diseases. Among the possible reasons:

1. Compost-treated plants are usually healthier and better able to resist infection, compost feeds microorganisms, which produce plant growth hormones and chelates that make micronutrients more available to plants, and
2. Compost hosts beneficial organisms that feed directly on disease organisms, compete with them for nutrients or produce antibiotics.

Some soils or potting mixes blended with medium- maturity compost - which still contains enough food for microorganisms - have sparked systemic resistance in plants, Hoitink says. 'These plants have elevated levels of biochemical activity relative to disease control and are better prepared to defend themselves against diseases.'

Not all composts provide this beneficial effect. In fact, composts and other biological materials that are rich in available nitrogen may actually stimulate some plant diseases. Among these diseases are *Phytophthora* root rot in soybeans, *Fusarium* wilts in vegetable crops and fire blight in fruit crops. To reduce the risk of initiating disease, spread these materials many months before cropping, allow the salts to leach away, or blend in low-nitrogen materials before application.

In Ohio, vegetable grower John Hirzel recorded 25-percent yield increases in tomatoes that were started in the greenhouse in mixtures of one-third compost, then transplanted to the field into soils amended with 10 to 12 tons of compost per acre. Hirzel, who died in 2000, found that tomatoes grown with more compost have better resistance to bacterial canker, bacterial spot and bacterial speck,. 'As soon as they germinate, they are living in a soil that has natural bacteria and fungi,' he said in a 1999 interview.

On the other hand, farming practices that cause imbalances in nutrition or other factors can lower natural resistance. High nitrogen fertilizer levels can fuel the germination and growth of many weed species, boost the incidence of diseases such as *Phytophthora, Fusarium* and corky spot, and stimulate outbreaks of aphids, mites and other insects.

Some herbicides lower the resistance of crops to invading disease-causing organisms. Even more serious, as it decays, glyphosate-treated vegetation can create flushes of *Fusarium, Rhizoctonia* and other pathogenic fungi.

Rotation, in the absence of known pests, has improved growth and yields in many crops by about 10 percent. Longer rotations tend to increase crop yields more than shorter rotations; yields of corn and wheat grown as part of three-year rotations exceed those in two-year cycles or in continuous monocultures. Adding organic matter - through cover cropping, animal manures and crop residues - boosts crop performance and may improve pest tolerance.

For three decades, Dick Thompson has planted cover crops, managed weeds like covers instead of like pests, and lengthened and expanded his crop rotation. 'I'm not saying we don't have any insect problems, but they do not constitute a crisis,' Thompson says. 'We don't have to treat for them. We haven't done that for years.'

Reduce Tillage for Healthy Soils

Many people consider tillage necessary for crop production. However, this seasonal practice can destroy some vital processes by depleting organic matter, intensifying the loss of topsoil to erosion and destroying soil tilth. Damaged soils are less able to provide nutrients, hold water and support biological activity. The net result: less diversity in crucial soil organisms.

'In my opinion, there's no single greater catastrophic event in the life of the soil than to have some big piece of tillage equipment run across it,' says Luna. 'Worms and bugs are killed, fungi are broken up and destroyed, and you end up with a much more simplified biological system.'

To plant and establish vigorous crops, you need to clear vegetation and residues from at least a portion of your field. Some equipment, however, minimizes soil disturbance. No-till planters, which cut a slot just wide enough to insert seed, disturb soil the least. Strip-tillage disrupts only a band of soil along the crop row, leaving untilled areas between rows. Ridge-till systems produce only shallow soil disturbances. Chisel plows do disturb soil structure, but, unlike moldboard plows, they do not invert or pulverize soils. No-till, zone-till and ridge-till also leave accumulations of plant residues covering the soil.

In a 1997 vegetable trial, Oregon State University researchers found a Willamette Valley farmer improved corn yields after strip-tilling into a winter cover of oats, vetch and Austrian winter peas. The farmer planted in eight-inch strips cut into the cover crop residue. The new system returned $100 per acre more than the standard tillage system.

Surface plant-residue mulches supply organic matter that reverses many of the detrimental effects of tillage. They take the edge off soil temperature extremes and keep soil moisture more consistent, thereby favoring a wide group of organisms. These factors combine to improve biological activity, soil tilth and nutrient- and water-holding capacity.

In Lancaster County, Pa., Groff says his cover crops - along with his 'full time, 100-percent commitment to no-till' - have increased his soil organic matter from 2.7 to 4 percent in the last decade. Although he farms on slopes as steep as 17 percent, his annual erosion losses are only a fraction of the county's average.

When putting in underground irrigation lines, Groff found 'roots of my rye cover crop 40 inches deep and earthworm holes 36 inches deep. By not tilling the soil, by leaving all of that structure intact, over several years I have a soil that begins to open up.'

Maintain Surface Residues

More diverse biological and physical environments at the soil surface spark more bountiful opportunities for regulating pests. The living and dead plant materials linked with no-till management readily establish biological activity, which can contribute to natural suppression of pests. The soil organic matter and fertility generated by cover cropping and reduced tillage also lessen pest damage simply by improving the growth and vigor of crops.

Cover crops supply generous amounts of surface vegetation and residue that can be customized for specific needs. Live, they furnish excellent habitat and food for beneficial insects. Strip-tilling a cash crop into a winter annual cover crop or overseeding the cash crop with a cover crop after the last cultivation allows bands of live covers to flourish between rows without over-competing.

In some systems, cover crops eliminate the need for pre-plant herbicides and reduce the need for post-emergence herbicides

Winter annual cover crops continue to yield crop benefits even after they have withered. Along with residue from previous crops, they can interfere with pest populations by:

- ✰ Hindering weeds or other soil pests by physically obstructing their growth, tampering with soil temperature or moisture, or unleashing plant-inhibiting allelopathic chemicals,
- ✰ Preventing fungal spores from being dispersed by water or wind, thereby curbing foliar diseases, and enhancing populations of predatory insects such as ground beetles and spiders.

'We see it over and over in our research,' says weed specialist Teasdale: 'The tomatoes in the vetch cover crop system stay green and healthier longer than tomatoes grown on a black plastic mulch.' Indeed, the tomatoes maintain healthy green foliage about 50 percent longer.

Other Practices for Building Healthy Soils

Good management of soil organic matter - reducing tillage, applying animal manures and composts, and rotating with such soil-building crops as sod-forming grass and legume forages - forms the basis for healthy soils. Develop strategies to encourage on-farm nutrient cycling and help organic matter accumulate. Take care to avoid compacting soils. You may need to keep heavy equipment off wet soils, maintain controlled traffic zones on soils susceptible to compaction, or use some tillage to break up compacted layers. Many of the soil-building practices discussed above help reduce soil erosion. On soils that are prone to erosion damage, consider strip cropping, grassed waterways to conduct runoff off fields, and soil terracing to help keep topsoil in place.

Create Multiple Stresses on Pests

Maximizing the impacts of many 'little hammers' takes an understanding of the life cycles of pests and of beneficial organisms. In ecologically based systems, farmers scrutinize the life cycles of pests and beneficial organisms, looking for times and places where small control measures can add up to big results. A good opportunity might arise during a pest's overwintering stage, for example, and another while it is first colonizing the crop. A beneficial organism may offer protection at one stage and need protection at another. Even in small increments, pest mortality can eventually pare a big problem to a low level.

In Sentinel Butte, N.D., cattle producer Dennis Dietz is battling a 75-80 percent leafy spurge infestation with several species of imported flea beetles after herbicides had little effect on a pest that plagues farmers in the western Plains. The flea beetles, however, are leaving their mark.

'I've seen dramatic changes in stem count and flowering,' says Dietz. 'My feeling is that the control is excellent, and it's long term.' Established in his own on-site insectary plantings, the flea beetles 'will be there forever,' he says. 'Chemical control, in my opinion, is way too expensive and it doesn't last as long.'

Not only do the adult beetles feed on the tops of leafy spurge, but flea beetle larvae burrow into the weed's roots, exposing them to a second stress: opportunistic disease organisms.

Cooperating in a 'Team Leafy Spurge' project through the USDA Agricultural Research Service in Sidney, Mont., Dietz has begun to add a third stress - sheep - to his biocontrol strategies. Their mission: graze off the tops of the plant while the beetles work on the roots.

Discourage the Pests' Dispersal or Connection with Crops

Interspersing non-host plants can hinder the movement of insect pests and crop disease organisms by altering light and humidity and jumbling the critical visual and chemical signals insect pests use to recognize their hosts. Non-host plants put distance between susceptible plants and, like fly paper, intercept spores to limit the spread of diseases.

Flea beetles that attack cabbage and other crucifers are less abundant when clover - a non-host species - is sown between cabbage rows. Crop losses to mildew are higher in pure stands of barley than in mixtures of barleys that differ in their susceptibility to disease races.

Disrupt Pest Populations by Destabilizing Habitat

When pests are adapted to some crops but not to others, rotations that include non-host crops can help with control. Placing a non-host crop in a rotation sequence often destroys the habitat a pest needs, limiting population growth. Infestations of Colorado potato beetles, for example, are more severe in continuous potatoes than in potatoes that follow winter wheat or rye. Similarly, rotating soybeans with such non-hosts as corn reduces charcoal rot.

To be suppressed by rotation, a target pest must have specialized feeding habits that restrict it to a narrow host range. Its ability to move to other locations also must be low. Finally, its dormant and resting stages must be shorter than the time gap between susceptible crops.

Rotations that include diverse crops and management practices tend to have fewer weed problems than simple rotations and crop monocultures. By continually changing the 'rules of the game,' complex rotations discourage the selection and adaptation of weeds.

Giant foxtail, for example, is less of a problem in a three-year corn-soybean-winter wheat rotation than in corn-soybean or continuous corn. Differences in the timing of germination, growth and competitiveness among the three crops, plus the suppressive effects of wheat straw, are likely reasons.

Reduce Weeds' Access to Resources

Many crops lack canopy cover and substantial root growth when they are young, limiting the plants' ability to fully capture sunlight, water and nutrients. Instead, these crops give invading weeds a foothold. Later in the season, weeds will compete with the crop and reduce its yield. Narrower row spacings, higher crop population densities and intercropping may all rob weeds of the resources they need to grow.

On the other hand, these same weed-choking practices can increase disease levels in some crops. Denser stands of wheat, for example, are more susceptible to mildew. Use disease-resistant varieties with such practices.

Enhance Beneficial Organisms

'Farmscaping' - a term coined by Robert Bugg of the University of California - describes a comprehensive approach to nurturing populations of beneficials. It examines and redesigns the whole farm landscape, rearranging fields, hedgerows, conservation buffers and other farm features to favor the beneficial organisms that protect crops.

Beneficial predatory and parasitic organisms generally do not flourish in fields with only one plant species. They need overwintering sites and different types of microenvironments - such as shady, moist places - where they can find protection from their own natural enemies. Besides the pests on which they prey, beneficials often need additional sources of food. Parasitic wasps and predacious hoverflies, for example, depend on a daily supply of honeydew, nectar and pollen for energy and reproduction. Alternative food sources are critical to the development of slow-reproducing predators.

To improve habitat for beneficials, consider:

Sowing cover crops between rows of cash crops,maintaining 'beneficial insectary plantings' at field edges, providing permanent refuge strips - 'or beetle banks' - for ground beetles, an important group of soil-dwelling generalist predators harboring natural predators, parasitoids and wildlife in perennial grasses, forbs, shrubs and trees on field edges or in strips, through conservation tillage, preserving soil structure and complex food webs for ground beetles and other beneficials, and supplying root disease-suppressing microbes with life-sustaining organic matter by means of cover crops, animal manures and composts.

'An intelligent addition to the diversity of habitat on the farm allows a lot of different kinds of predators and parasites to work on the side of the farmer,' says Kim Stoner, assistant entomologist at the Connecticut Agricultural Experiment Station in New Haven. 'It makes sense to go for a spectrum of flowering plants over the course of the year.'

In western Texas, pecan grower Kyle Brookshier uses one key strategy to limit dispersal of stink bugs. He plants black-eyed peas between or around all 1,300 acres of his nut trees. Drawn to the peas, the stink bugs now leave his pecans virtually alone. Rather than the 12–13 percent damage they used to cause in his nut crop, Brookshier now sees less than 1 percent.

'We have almost ceased to get damage from stink bug,' he says. 'I think it should be a standard cultural practice in pecan orchards.'

Brookshier plants the peas two to four rows wide at two-week intervals between late June and late July. That way, his trap crop is always lush when the stink bugs are active. An added bonus of this inexpensive pest control strategy: The black-eyed peas are a hit at family meals.

Only as a Last Resort, Use Targeted Attacks

Even in ecologically based pest management systems, farmers may need to use pest control tactics, including pesticides. Managing weeds without tillage or herbicides, for example, is not consistently reliable. Because unwanted populations of annual and perennial vegetation can build very rapidly, herbicides remain an important tool, especially in no-till systems.

Judicious selection and limited use of herbicides that are low in toxicity and short in environmental persistence - combined with minimum-till and cover crop management - will help create habitat for beneficial organisms and develop healthy soils.

Sometimes key insect and disease pests - often introduced from another part of the world - can damage crops significantly. Ecologically based controls may not be available for these recently imported species. In this situation, reacting with the least disruptive, most specific chemical may be the farmer's best option.

Use reactive interventions only after clear decision-making. As you assess, consider the following:

properly identifying the pest and possible beneficial species present, assessing the pest population and its threat to the crop, and selecting the appropriate tactic - a chemical, biocontrol organism or other intervention - based on full knowledge of the range of measures available and their effectiveness, cost and side effects.

Options for Pesticides and Biorationals

To kill pests, disrupt their life cycles or deny their access to crops, farmers have an assortment of conventional and biorational materials at their disposal. Conventional chemicals include synthetic, broad-spectrum pesticides that often leave in their wake unwanted side effects - harming other species or polluting the environment.

Biorationals are more specifically toxic to or disruptive of target pests. Naturally derived or synthesized, they include growth regulators, microbial toxins, anti-feeding agents, pest-smothering oils, and disruption pheromones that confuse insects and reduce their reproductive success.

'They're an improvement and - if used properly - there should be an economic gain to the grower,' says Ed Rajotte, integrated pest management coordinator at Penn State University. For now, he says, many biorationals are more expensive and more difficult to use.

Rajotte's emphasis today: helping farmers substitute the many 'little hammers' of management information for the 'big hammer' of broad-spectrum pesticides.

If the agricultural research and extension community applies 'a concerted effort' over the next decade, Magdoff believes ecologically based pest management systems could be widely adopted.

'A lot of people have parts of this ecologically based pest management system working very well for them right now,' says Magdoff.

Beware the temptations of the 'big hammer,' says Fred Kirschenmann, an organic grain farmer in North Dakota. Like everyone, he points out, farmers want to see

immediate results. Quick satisfaction from a big hammer strategy often gives way to disappointment over the long term.

'Develop the attitude that every time a 'big hammer' strategy is used, it represents a failure in the system,' Kirschenmann says. 'You should always assess what went wrong and what strategies to follow up with to put the 'many little hammers' back in place.'

Index

A

Abiotic components, 3
Abundance, 57, 58, 75
Acceptable pest levels, 161
Adalia bpunctata, 64
Adhesion, 122
Aestivation, 25, 57, 72
Agricultural ecosystem, 84
Allelochemicals, 13
Allelopathy, 13
Amsacta spp, 26
Alpha diversity, 17
Allomones, 118
Amensalism, 90
Amphibians, 73
Amphibian populations, 164
Analogous, 111
Ancestral trait, 112
animal communities, 2
Animal ecology, 4
Anomala dimidiata Hope, 154
Antagonism, 90
Antagonistic interaction, 90
Antartica, 16
Antipredator adaptation, 125
Apis laboriosa Smith, 154
Aphilinus mali, 98
Apomorphies, 112
Aposematic, 115
Aposematic coloration, 123
Aposematism, 115
Apparent competition, 87
Applied ecology, 2
Aquarius paludum, 126
Arbuscular mycorrhizal fungi, 158
Arthropods, 16
Aretogaea, 17
Aspection, 110
Arphia sulphurea, 110
Arphia xanthoptera, 110
Assimilation rate, 2
Augmentation, 163
Aulacophora foveicollis Lucas, 27
Autecology, 6
Autotomy, 117

B

Bacillus papillae, 61
Bacillus thurenginensis, 61
Batesian mimicry, 117, 124
Beetle banks', 176
Behavioural neuroscience, 6
Bemasia tabaci, 66
Beta diversity, 18
Big hammer', 177
Biochemical reactions, 3
Bioclimate range, 13
Biocoenosis 2, 8, 31, 35
Biogeography, 11
Biological Diversity, 3
Biological magnification, 82, 83
Biological spectrum, 6
Bioluminescence, 28
Biomass, 82
Biorationals, 177
Biospheres, 7, 8
Biotic principles, 1
Birth rate 40, 40
Bombyx mori, 28
Boysenberries, 169
Broomistega putterilli, 73
Brookshier plants, 176
Browse, 130
Bt hybrid, 166

C

Camouflage, 115, 123
Carnivores, 12, 75
Carrying capacity, 44
Caverns, 16
Central Limit Theorem, 138
Chisel plows, 173
Chorthipus sp, 154
Cladistics, 112
Cleorodes lichenaria, 126
Climatic factors, 2
Climax communities, 2
Clumped distribution, 12
Coccinella undecimpunctata, 64
Cognition, 9
Co enzymes, 78
Commensalism, 76, 91
Communities, 1
Community ecology, 2, 129
Community equilibrium, 2
Community gardens, 162
Community types, 5
Competition, 2
Competition coefficient, 102
Competitive ability, 2
Competitive exclusion, 2
Competitor, 3
Components, 5
Conservation ecology, 6
Consequenses, 87
Consumers, 2
Cost benefit ratio, 92
Coundershaded, 116
Cover crops, 173
Coyote, 14
Crepuscular, 28
Cross-ecosystem, 158
Crustacea, 16
Cry protein, 165
Crypsis, 123
Cryptochaetum iceryae, 95
Cylindrocarpon, 171
Cytoecology, 6

D

Darwin's Theory, 3
Degree days, 161
Dendroctonus frontalis, 68
Developmental zero, 24
Diabrotica virgifera virgifera LeConte, 166
Diapause, 57
Discomfort, 122
Discrete model, 37, 40, 41
Dispersal, 10, 25, 59
Diurnal, 27
Diurnation, 110
Divergent, 114
Diversity stability debate, 155
Diversity–stability hypotheses, 158
Danaus plexippus, 154
Drier season, 70
Dynamics, 2

E

Earias fabia Stoll, 27
Ecdysone precursor, 77
Ecological Animal Geography, 2
ecological niches, 2, 109
Ecological facilitation, 91
Ecological homologous, 111
Ecological succession, 81
Ecologists, 1
Economic distribution, 16
Ecosystem, 2
Ectoparasite, 62
Eilema spp, 126
Endoparasite, 62
Energy budgets, 2
Energy flow, 3
Entomopathogenic nematodes, 162
Entomopathogenic fungi, 162
Entomophagous, 76
Environment3, 5
Environmental resistance, 45
Ephydra cinera, 16
Equilibrium dynamics, 157
Equilibrium resilience, 157
Eretomocerus serius, 98
Eurythermal-insects, 16
Exponentail model, 37, 38, 39

F

Facultative, 71
Fecundity, 25, 27, 44
Folivore, 130
Food chains, 2
Freeze avoidance, 74
Freeze tolerance, 74
Frugivores, 125
Fundamental niche, 109
Fusarium wilts, 172

G

Gala hypothesis, 3
Galton Watson model, 139
Gamma diversity, 19
genetic equilibrium, 2
Genomes, 8
Geographical race, 34
Giant foxtail, 175
Glyphosate, 172
Grazers, 130
Green Pest Management, 164
Guild, 129

H

Habitat ecology, 6
Haplodiploid inheritance, 128
Helaeomyia petrolei, 16
Herbivorous, 75
Hibernation, 25, 57, 73
Hierarchy of life, 7, 8

Hippodamia convergens, 64
Holism, 8
Homoeothermic, 23
Homogeneity, 19
Homoplasious traits, 112

I

Immunity, 121
Individual, 3
Individualistic, 132
Inheritance, 2, 111
Insect ecology, 5
Inter specific competition, 2, 87
Intra specific competition, 2
Intra specific interaction, 85
Iteroparous, 41
Iteroparity, 41

J

Juvenile hormone analogues, 162

K

Kin selection, 120

L

Latitudes, 19
Latitudinal gradient, 20
Law of minimum, 31
law of tolerance, 2, 31
Leucaena leucocephala, 13
Libellula, 64
Life table, 51, 52, 53
Limnology, 2
Little hammers', 174, 177
Livestock, 5
Logistic model, 37, 38
Logos, 3, 5
Log series distribution, 137
Lotka–Volterra equations, 101
Lycaon pictus, 12

M

Marion berries, 169
Masquerade, 115
Medauroidea extradentata, 116
Metabolisms, 3
Metacommunity, 141
Microbial interactions, 158
Micro climate, 110
Microcosm, 2
Micro environments, 175
Micro habitat, 110
Mimesis, 115, 123
Mimic, 115
Mimicry 109, 115, 124
Molding, 6
Monitoring, 161
Monoculture, 169
Monophagous, 86
Monophyletic clade, 112
Müllerian mimicry, 117
Multitrophic, 158
Mutalism, 86

N

Natural selection, 2, 3, 5
Natural world, 4
Nemoria darwiniata, *121*
Neogaea, 17
Nephotettix spp, 27
Neutralism, 90
Niche apportionment model, 137
Niche differentiation, 88
Niche-pre-emption model, 137
Nilaparvata lugens, 94, 164
Nocturnal, 27
No-till cropping system, 168
Notogaea, 17
Nutrient budget, 3

O

Obligatory, 71
Offspring, 3, 11
Oikos, 3, 5
Oligophagous, 75
Omnivorous, 75
Oogenesis flight syndrome, 67
Organelle, 7
Organization, 7
Organisms, 3
Origin of Species, 3, 133
Ornithoptera alexandrae, 154
Oscillatory dynamics, 159
Ostrinia nubilalis Hübner, 166
Otala lactea, 72
Oviposition, 28
Oyster beds, 2

P

Paleoecology, 6
Pantala, 64
Paraphyletic species, 112
Parasite, 61, 63
Parasitoids, 62
Parnassius Mnemosyne, 126
Pathogens, 61
Pectinophora gossypiella, 28
Perennial plants, 168
Periodic cycle, 86
Persistence, 65
Pestiferous, 171
Pharmacology, 7
Phonological models, 166
Phenotypic plasticity, 121
Pheromones, 118
Photoperiod, 71
Photoperiodism, 28
Phototaxis, 27
Phragmosis, 120
Phylogenetic clades, 12
Physical attributes, 3
Physical surroundings, 3
Phytophagous, 76
phytoplankton communities, 156
Phytophthora root rot, 172
Pieris canidia indica Sparr, 154
Pieris napi, 126
Pigmentation, 28
Pilocity, 26
Planktonic copepods, 133
Plesiomorphy, 112
Poikilothermic, 23
Poikilothermous, 16
Poisson distribution, 15
Polymorphism, 125
Polyohagous, 75
Population, 3
Population biologist, 2
Population dynamics, 2
Population ecology, 2
Population Genetics, 2
Population size, 2
Predation, 2
Predator satiation, 125
Predators, 5, 61
Pre-emergent herbicide, 163
Preventive conservation, 161
Prey- predators interactions, 115
Primary defense, 115
Producers, 2
Production ecology, 6
Prospatella berlesei, 98
Protective coloration, 122
Prothoracicotrophic hormone, 71
Pythium, 171

Q

Quiescence, 57

R

Rapid development, 3
Realized niche, 109
Reductionism, 8
Reductionistic approach, 7
Regurgitation, 120
Repellency, 122
Reproductive rates, 3
Rhizoctonia, 171, 172
Risk assessment, 163, 164
Rodalia cordinalis, 95

S

Salvia leucophylla, 13
Saprophagous, 76
Scavenger, 76
Scirpophaga incertulas, 165
Scramble competition, 87
Semelparity, 41
Semelparous, 41
Serengeti community dynamics, 158
Shannon–Wiener Diversity 1, 50
Shared habitat, 3
Shared system, 3
Socio-homoeothermic, 24
Spatial dimension, 166
Spatial maps, 15
Species, 3
Species niche, 2
Species range, 10
Sørensen index, 18
Stenothermal-insects, 16
Sterile insect technique, 163
Strip cropping, 174
Strip-tillage, 173
Succession, 1
Supervised insect control, 160
Survivors, 3
Survivorships curve, 49, 50, 51
Symbolism, 153
Symbiosis, 76
Sympetrum, 64
Synecology, 6
Synergistic action, 32
Syrphus fulvifacies Brunetti, 154
Systemic acquired resistance, 171
System ecology, 35

T

Telonomus beneficiaens, 76
Temperature preferendum, 25
Terpenes, 13
Territoriality, 13, 16
Thermal constant, 25
Transgenic Agro ecosystem, 165
Transportation, 10
Tribolium casteneum, 61
Tribolium confusum, 61
Trolocha virescence, 53, 54, 55
Trophic levels, 2, 82

U

Umbrella effect, 153

V

Vanessa cashmirensis Fru, 154
Variance, 15
Venom, 119
Vito Volterra, 2

W

Weakness, 21

Y

Yule model, 139

Z

Zone tillage, 168
Zoophagous, 76